**Experimental Investigation of Turbulent Boundary Layer with Uniform Blowing at Moderate and High Reynolds Numbers**

# Experimental Investigation of Turbulent Boundary Layer with Uniform Blowing at Moderate and High Reynolds Numbers

Von der Fakultät für Maschinenbau, Elektro- und Energiesysteme der Brandenburgischen Technischen Universität Cottbus–Senftenberg zur Erlangung des akademischen Grades eines Doktor der Ingenieurwissenschaften (Dr. -Ing.)

genehmigte Dissertation
vorgelegt von
M. Sc. Gazi Hasanuzzaman
geboren am 15. November 1984 in Dhaka, Bangladesh

| | |
|---|---|
| Vorsitzender: | Prof. Dr. rer. nat. Andreas Schröder |
| Gutachter: | Prof. Dr. -Ing Christoph Egbers |
| Gutachter: | Prof. Dr. -Ing. Jean-Marc Foucaut |
| Tag der mündlichen Prüfung: | 05. Mai 2021 |

**Bibliografische Information der Deutschen Nationalbibliothek**
Die Deutsche Nationalbibliothek verzeichnet diese Publikation in der Deutschen Nationalbibliografie; detaillierte bibliographische Daten sind im Internet über http://dnb.d-nb.de abrufbar.
1. Aufl. - Göttingen: Cuvillier, 2022
Zugl.: (TU) Cottbus-Senftenberg, Univ., Diss., 2021


Nonnenstieg 8, 37075 Göttingen
Telefon: 0551-54724-0
Telefax: 0551-54724-21
www.cuvillier.de


1. Auflage, 2022
Gedruckt auf umweltfreundlichem, säurefreiem Papier aus nachhaltiger Forstwirtschaft.

ISBN 978-3-7369-7558-3
eISBN 978-3-7369-6558-4

# Acknowledgment

In order to complete present research, biggest stakeholders of all are:

1. Department of Aerodynamics and Fluid mechanics
2. Graduate Research School: SP 4 - Flow control of a flat plate turbulent boundary layer flow through micro blowing under stochastic forcing.
3. Laboratoire de Mécanique de Feiret Lille from Université de Lille

I would like to cordially thank my supervisor Prof. Dr. -Ing. Christoph Egbers to believe in me since I have started working with him during my masters thesis., subsequently, for his opportunity of undertaking the doctoral studies under his guidance. His benevolent support and reliance on me, helped significantly while working on this project.

My sincere gratitude goes to Dr. -Phys. Sebastian Merbold, whos visionary supervision and knowledge in the course of this thesis has provided pace and bolstered every individual milestones set for the project.

I sincerely appreciate the supportive conversations with Prof. Dr. El-Sayed Zanoun, Prof. Dr. Uwe Harlander, Dr. Vasyl Motuz, Stefan Richter and Dr. -Ing. Andreas Froitzheim which has helped me gain insight about the topic.

I convey my sincere thanks to the financial and administrative support of Mr. Robert Rode from Graduate Research School which has helped a lot realizing this project.

Particularly, for the experiment presented in Chapter-3, I would like to acknowledge the financial support of the European High-performance Infrastructure in Turbulence (EuHIT) under the project name "Enhanced Turbulent Outer Peak using uniform Micro-Blowing Technique" within the frame of its Transnational Access activity. This is a Project coordinated by Max Plank Institute Göttingen and funded by the European Commission under Grant agreement no: 312778, in the frame of the Research Infrastructures Integrating Activity (framework of FP7-INFRASTRUCTURES-2012-1). In particular Dr. -Ing. habil. Holger Nobach for his financial and organizational guidance.

I would be grateful to our partner institution Laboratoire de Mécanique des Fluides de Lille, specially Prof. Jean-Marc Foucaut and Dr. Christophe Cuvier for their benevolent approval and support while conducting measurement at their boundary layer wind tunnel. The Laboratoire de Mécanique des Fluides de Lille equipment (the wind-tunnel, the PIV equipment and the compressed air regulation and quantification circuit) used during this experiment was funded by the ELSAT2020 project supported by the European Community, the French Ministry of Higher Education and Research and the Hauts de France Regional Council, in the framework of the CNRS Research Foundation on Ground Transport and Mobility.

Last but not the least, I am grateful to my beloved wife Nazmoon Nahar Munni and my daughters Fatima Gazi and Aymaa Gazi, who remained patient and continuously supported my sleepless nights and busy days during the course of this thesis. I am forever indebted to my parents A. K. M. Asaduzzaman and Nasrin Akter, for their selfless contribution that helped me reach here today. Finally, I would be ever grateful to almighty God for the blessings he has bestowed upon me with all his mercy and grace.

# Abstract

Experimental investigation in turbulent boundary layer flows represents one of the canonical geometries of wall bounded shear flows. Utmost relevance of such experiments, however, is applied in the engineering applications in aerospace and marine industries. In particular, continuous effort is being imparted to explore the underlying physics of the flow in order to develop models for numerical tools and to achieve flow control. Flow control experiments have been widely investigated since 1930's. Several flow control technique has been explored and have shown potential benefit. But the choice of control technique depends largely on the boundary condition and the type of application. Hence, friction drag of subsonic transport aircraft is intended to be reduced within the scope of this Ph. D. topic. Therefore, application of active control method such as micro-blowing effect in the incompressible, zero pressure gradient turbulent boundary layer was investigated. A series of experiments have been performed in two different wind tunnel facilities. Wind tunnel from Department of Aerodynamics and Fluid Mechanics (LAS) was used for the measurements for moderate Reynolds number range in co-operation with the wind tunnel from Laboratoire de Mécanique de Feiret Lille for large Reynolds number range. Measurements are conducted with the help of state-of-the-art techniques such as Laser Doppler Anemometry, Particle Image Velocimetry and electronic pressure sensors.

Present control experiments in turbulent boundary layer can be split into two different work segments, where one is objected towards the data measurements in turbulent boundary layer over smooth surface with and without any external perturbation. Here, perturbation is applied in the form of wall normal blowing while keeping the magnitude of blowing very low compared to the free stream velocity. For the subsequent results reported here, magnitude of blowing ratio was varied between 0% ~ 6%.

Turbulent boundary layer flow is particularly interesting as well as challenging due to the presence of different interacting scales which are increasingly becoming significant as the flow inertial conditions keeps growing. Therefore, energy content of the coherent structures in outer layer becomes stronger and necessitates measurements in relatively large Reynolds number. Hence, a set of LDA and PIV measurements have been performed in the range of shear Reynolds number within $0.415 \times 10^3 \leq Re_\tau \leq 5.616 \times 10^3$.

In the first part of the present thesis e.g $0.415\times10^3 \leq Re_\tau \leq 1.160\times10^3$, measurements were performed at the Brandenburg University of Technology wind tunnel. Non-intrusive Laser Doppler Anemometry was applied to carry out a series of measurements on a zero pressure gradient flat plate turbulent boundary layer. Smooth and perforated surfaces were used for 16 different momentum thickness Reynolds numbers ranging from 1100 to 3700. Blowing ratio through the perforated surface was varied between 0.17% ~ 1.52% of the free stream velocity. To a maximum of 50% reduction in friction drag was achieved, a quantitative analysis of friction coefficient as a function of flow Reynolds number is presented and discussed. Additionally, boundary layer thickness increases as the amplitude of blowing air increases. Corresponding Mean profiles of streamwise velocity and normalized statistics of fluctuating data is found to be in good agreement with numerical data. As a consequence of blowing, gradient of the streamwise fluctuation is strongly influenced in contrast to the first peak.

For the measurements on the upper range of the stated Reynolds number, were conducted at the boundary layer wind tunnel at Laboratoire de Mécanique des Fluides de Lille. This boundary layer wind tunnel offers a spatially developed turbulent boundary layer over a flat plate within $2.2 \times 10^3 \leq Re_\tau \leq 5.5 \times 10^3$ with an excellent spatial resolution. With the help of state-of-the-art 3D Stereo Particle Image Velocimetry technique, high fidelity measurement of the velocity components were obtained covering entire boundary layer in streamwise wall normal plane. In addition, time resolved measurements were also obtained in spanwise and wall-normal plane in order to look into the morphology of turbulent structures immediately above the blowing area.

# Zusammenfassung

Experimentelle Untersuchungen in turbulenten Grenzschichtströmungen stellen eine der kanonischen Geometrien wandgebundener Scherströmungen dar. Die größte Relevanz solcher Experimente liegt in den ingenieurtechnischen Anwendungen in der Luft- und Raumfahrt sowie in der Schifffahrt. Insbesondere wird kontinuierlich daran gearbeitet, die zugrunde liegende Physik der Strömung zu erforschen, um numerische Modelle zu entwickeln und eine Strömungssteuerung zu ermöglichen. Experimente zur Strömungskontrolle wurden seit den 1930er Jahren umfassend untersucht. Verschiedene Techniken zur Strömungskontrolle wurden erforscht und haben ihren potenziellen Nutzen gezeigt. Aber die Wahl der Kontrolltechnik hängt weitgehend von den Randbedingungen und der Art der Anwendung ab. Im Rahmen dieses Promotionsthemas soll der Reibungswiderstand von Unterschall-Transportflugzeugen reduziert werden. Daher wurde die Anwendung einer aktiven Regelungsmethode, wie z.B. dem Effekt des Mikro-Ausblasens, in der inkompressiblen turbulenten Grenzschicht mit Nulldruckgradienten untersucht. Eine Reihe von Experimenten wurde in zwei verschiedenen Windkanälen durchgeführt. Für die Messungen wurde der Windkanal des Lehrstuhls für Aerodynamik und Strömungsmechanik (LAS) für den mittleren Reynoldszahlenbereich genutzt, in Zusammenarbeit mit dem Laboratoire de Mécanique de Feiret Lille (LMFL) der Windkanal für den großen Reynoldszahlenbereich. Die Messungen werden mit Hilfe modernster Techniken wie der Laser Doppler Anemometrie, Particle Image Velocimetry und elektronischen Drucksensoren durchgeführt.

Heutige Kontrollexperimente in der turbulenten Grenzschicht können in zwei verschiedene Arbeitsbereiche aufgeteilt werden, wobei der eine auf die Datenmessungen in der turbulenten Grenzschicht über glatter Oberfläche mit und ohne externe Störung gerichtet ist. Hier wird die Störung in Form von orthogonaler Wandanblasung erzeugt, wobei die Größe der Anblasung im Vergleich zur freien Strömungsgeschwindigkeit sehr gering gehalten wird. Für die hier gezeigten Ergebnisse wurde die Größe des Anblasens zwischen 0% ~ 6% variiert.

Die turbulente Grenzschichtströmung ist besonders interessant und herausfordernd aufgrund der Anwesenheit von verschiedenen interagierenden Skalen, die zunehmend an Bedeutung gewinnen, wenn die Trägheitsbedingungen der Strömung weiter zunehmen. Daher wird der Energiegehalt der kohärenten Strukturen in der äußeren Schicht stärker und macht Messungen bei relativ groß en Reynoldszahlen erforderlich. Aus diesem Grund wurde eine Reihe von LDA- und PIV-Messungen im Bereich der Scher-Reynoldszahl von $0,415 \times 10^3 \leq Re_\tau \leq 5,616 \times 10^3$ durchgeführt.

Im ersten Teil der vorliegenden Arbeit werden Messungen im BTU-Windkanal vorgestellt, welche beispielsweise für einen Bereich von $0,415\times10^3 \leq Re_\tau \leq 1,160\times10^3$ durchgeführt wurden. Mit der nicht-intrusiven Laser-Doppler-Anemometrie wurde eine Messreihe an einer turbulenten Plattengrenzschicht mit Nulldruckgradienten durchgeführt. Es wurden glatte und perforierte Oberflächen für 16 verschiedene impulsdicken-bezogene Reynoldszahlen im Bereich von 1100 bis 3700 verwendet. Das Geschwindigkeitsverhältnis der durch die perforierte Oberfläche geblasenen Strömung zur freien Anströmung wurde zwischen 0,17 % ~ 1,52 % variiert. Es wurde eine maximale Verringerung des Reibungswiderstandes von 50% erreicht. Eine quantitative Analyse des Reibungskoeffizienten als Funktion der Strömungs-Reynoldszahl wird vorgestellt und diskutiert. Zusätzlich nimmt die Grenzschichtdicke mit zunehmender Stärke des eingeblasenen Luftstromes zu. Entsprechende Geschwindigkeitsprofile in Strömungsrichtung und normalisierte Statistiken der fluktuierenden Daten zeigen eine gute übereinstimmung mit den numerischen Daten. Als Folge des Anblasens wird der Gradient der Fluktuation in Strömungsrichtung im Gegensatz zum ersten Peak stark beeinflusst.

Die Messungen im oberen Bereich der angegebenen Reynoldszahl wurden im Grenzschichtwindkanal des LMFL durchgeführt. Dieser Grenzschichtwindkanal bietet eine räumlich entwickelte und hervorragend aufgelöste turbulente Grenzschicht über einer ebenen Platte in einem Bereich von $2,2 \times 10^3 \leq Re_\tau \leq 5,5 \times 10^3$. Mit Hilfe der hochmodernen 3D Stereo Particle Image Velocimetry Technik wurden High-Fidelity-Messungen der Geschwindigkeitskomponenten über die gesamte Grenzschicht in der strömungsseitigen Wandnormalen-Ebene gemacht. Darüber hinaus wurden auch zeitaufgelöste Messungen in der Spannweiten- und Wandnormalebene durchgeführt, um die Morphologie der turbulenten Strukturen unmittelbar über dem Einblasbereich zu untersuchen.

# Other Publications

The following papers are included in this thesis.

1. Hassanuzzaman, G., Merbold, S., Motuz, V. and Egbers, Ch., (2021). Enhanced outer peaks in turbulent boundary layer using uniform blowing at moderate Reynolds number, Jour. Turb. , DOI:10.1080/14685248.2021.2014058.

2. Hassanuzzaman, G., Merbold, S., Cuvier, C., Motuz, V., Foucaut, J.-M., and Egbers, Ch.,(2020).Experimental investigation of turbulent boundary layers at high Reynolds number with uniform blowing, part I: statistics, Jour. Turb., DOI: 10.1080/ 14685248. 2020. 1740239, 21(3), pp. 129 – 165.

The following papers, although related, is not included in this thesis.

1. Hassanuzzaman, G., Merbold, S., Motuz, V., Egbers, Ch., Cuvier, C. and Foucaut, J-M.(2018).Experimental investigation of active control inturbulent boundary layer using uni-form blowing,5th International Conference on Experimental Fluid Mechanics (ICEFM), 2.- 4. July 2018, pp. 1 – 6.

2. Hassanuzzaman, G., Merbold, S., Motuz, V., Egbers, Ch., (2016).Experimental Investigation of Turbulent Structures and their Control in Boundary Layer Flow, Fachtagung Experimentelle Strömungsmechanik, 6. - 8. September 2016, pp. 64:1 – 6.

# Other Publications

# Contents

# List of Figures

# List of Tables

# Nomenclature

$C\,[m]$ ,Chord length of an airfoil

**Greek symbols**

$\delta\,[m]$ ,boundary layer thickness is the distance from wall where mean streamwise velocity reaches 99% of free stream velocity.

$\delta^*\,[m]$ ,displacement thickness is defined as the distance by which the external potential flow is displaced outwards due to the decrease in streamwise velocity of the boundary layer profile

$\mu = \rho\nu\,[Ns/m^2]$ ,dynamic viscosity

$\nu\,[m^2/s]$ kinematic viscosity

$\rho\,[kg/m^3]$ density of the fluid

$\tau_W = \mu d\bar{u}/dy\,[kg/ms^2]$ ,Mean streamwise wall shear stress representing the shear force per unit area exchanged between the surface and the fluid

$\theta\,[m]$ ,momentum thickness is defined as the loss of momentum in the boundary layer as compared that of the potential flow.

**Other Symbols**

$1/2\rho U_\infty^2\,[N/m^2]]$ ,free stream dynamic pressure

$\text{BR}\ [\%] = \frac{V_W}{U_\infty} \times 100$ ,Blowing ratio

$H = \frac{\delta^*}{\theta}$ ,Shape factor is the ratio between $\delta^*$ and $\theta$

$M = \frac{V}{c}$ ,Mach number is a quantity that defines the velocity of the flow (V) with respect to the velocity of sound (c)

$Re_\tau = \frac{u_\tau \delta}{\nu}$ ,shear Reynolds number

$Re_\theta = \frac{U_\infty \theta}{\nu}$ ,Reynolds number based on momentum thickness

$Re_C = U_\infty C/\nu$ ,chord Reynolds number

$Re_x = \frac{U_\infty x}{\nu}$ ,Reynolds number based on characteristics length

$Re_{\delta^*} = \frac{U_\infty \delta^*}{\nu}$ ,Reynolds number based on displacement thickness

$u_\tau = \sqrt{(\tau_w(x)/\rho)}\,[m/s]$ ,Local wall shear velocity

$V_W$[m/s] ,Applied blowing velocity in wall normal direction

**Physics symbols**

$C_f = \tau_W/0.5\rho U_\infty^2$ ,Skin friction co-efficient is a dimensionless skin shear stress which is non-dimensionalized by the dynamic pressure of a free stream

$U_\infty\,[m/s]$ ,time averaged free stream velocity

$u_{rms}$ or $\sqrt{\overline{u'^2}}$ [m/s] , root-mean-square of streamwise velocity fluctuation

$v_{rms}$ or $\sqrt{\overline{v'^2}}$ [m/s] , root-mean-square of wall-normal velocity fluctuation

$w_{rms}$ or $\sqrt{\overline{w'^2}}$ [m/s] , root-mean-square of spanwise velocity fluctuation

# 1 Introduction

"A very satisfactory explanation of the physical process in the boundary layer between a fluid and a solid body could be obtained by the hypothesis of an adhesion of the fluid to the walls, that is, by the hypothesis of a zero relative velocity between fluid and wall. If the viscosity was very small and the fluid path along the wall not too long, the fluid's velocity ought to resume its normal value at a very short distance from the wall. In the thin transition layer, however, the sharp changes of velocity, even with small coefficient of friction, produce marked results."

- Ludwig Prandtl (1904) – Address to the 3rd Mathematical Congress in Heidelberg

## 1.1 Motivation

Our understanding of TBL over flat plate related to its function in flight, heat convection in turbines or regarding propellers, hydrodynamics and several other similar engineering applications had been immensely investigated during last century and has continues till today. In a more general sense the wall-bounded turbulent flows are of special importance due to their scientific relevance and also their technical applications. Therefore, they have been regarded as an important topic for physicists, mathematicians and engineers, inspiring wide ranges of studies. The intrigue boundary conditions imposed by the wall bounded turbulent flows in terms of the structures and their scales in the process of transport and dissipation of the turbulent kinetic energy, plays a vital role determining associated friction drag. On the other hand, skin friction drag as a function of total drag, is of keen interest in terms of practical applications.

Flow Control Technique's for turbulent boundary layer has an wide engineering application from modern aircraft to the high speed trains and cars. Estimations show that, nearly 50% of the total drag of an subsonic aircraft and 30% from automobiles are constituted from skin friction drag (Kornilov (2005)). Airline and truck industry consumes 238.5 billion and 190 billion litres of oil per year respectively, out of which 25% and 27% fuel is spent to overcome the viscous drag (Wood (2004)). In United States alone, 40% drag is coming from skin friction in transportation sector whereas for a subsonic long range passenger liner, approximately 50% of the total drag is contributed from friction, 70% for marine vessels and ~100% for long pipeline pumping (Wood (2004)). Only 1% of reduction of such skin friction drag can save upto 400,000 litres of fuel yearly (based on Airbus A320 aircraft data (Kim et al. (2011))).

Large part of the fuselage, wing, tail wing and radar section of a subsonic aircraft has the potential to deploy drag reduction mechanisms. Due to their size and operating speeds, the majority of commercial and military aircraft in service today are dominated by flows that results from the presence of turbulent boundary layer. This generally cover the most of the aircraft's surface. It is well known that turbulent boundary layer significantly increases the skin friction drag penalties when compared to laminar boundary layers. Moreover, they do result in a reduced susceptibility to flow separation due to their robustness to surface imperfections. Therefore, turbulent drag reduction has a direct relationship to the eddy structures of different size and scales present in the boundary layer.

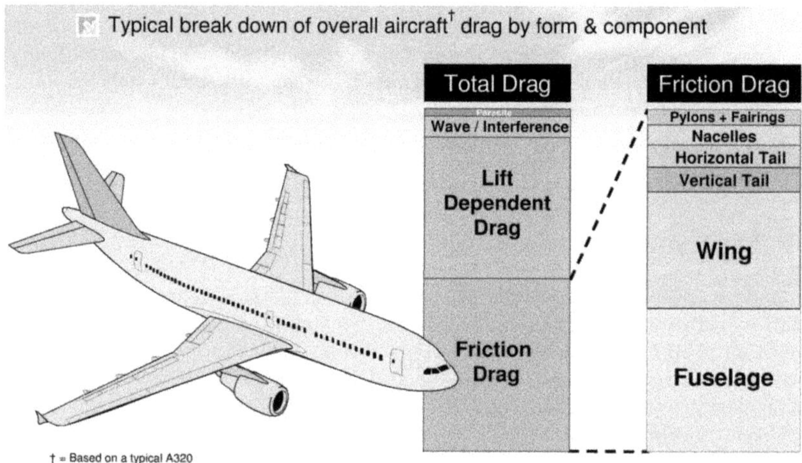


**Figure 1.1:** Break down of the drag contribution according to the drag types on different aircraft component (Hills (2008)).

Form/Pressure drag contribution to an aerodynamic body can be minimized with an optimized and streamlined design where as skin friction drag being the largest contributor than the former one. Apart from advances in designing streamlined body for aircraft, reduction of friction drag is the only possibility to improve aerodynamic efficiency (Kim et al. (2011)).

Figure-1.1 exhibit the contribution of different types of drag on a subsonic A320 aircraft according to its surface sections. This indicates that nearly half of the total drag comes from the friction drag. For aero and thermodynamic applications, flow control based on the varying surface conditions has been and still, of keen interest where motivation is focused towards the drag reduction, which in turn, can lead to significant financial benefit.

## 1.2 Turbulent flows

Most of the flows we encounter in nature as well as engineering applications are turbulent. It is both spatial and temporal instability and swirling of different scales caused by a continuous mixture of complex flows. A turbulent flow may arise due to the effect of frictional force on the surface of a stationary wall being flowed around or due to the interaction between flows at different rates. Moreover, a turbulent flow is characterized by the fact that its velocity and pressure in a spatial reference point are unstable over time. Particularly for wall bounded turbulent flows, the bounding energy content between the free stream flow and the underlying surface fluctuate depending on the flow Reynolds number and viscous effect. This type of instability with continuous change of state means that a turbulent flow can only be characterized by a series of parameters. These parameters include:

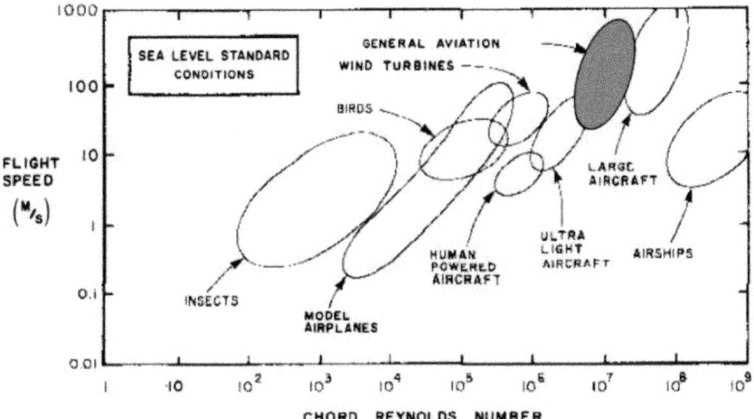


**Figure 1.2:** Operating range of different flying objects where red oval indicate the ranges for the general aviation (Lissaman (1983)).

1. Unsteadyness: A turbulent flow is irregular, disordered or chaotic and random. Despite chaotic behavior, the turbulence is de-terminus and can be described by the Navier-Stokes equations. Many probabilistic turbulence models allow to determine the most important parameters of the turbulent flow such as speed, temperature and pressure.

2. Diffusion: Increase in diffusion also causes an increase of the wall shear stress in the near wall region and results in an increase of the frictional resistance.

3. Dissipation: The kinetic energy of the flow is very quickly converted into internal energy of the small vortex or turbulence scales.

4. Three-dimensionality: The turbulent flow is always three-dimensional. It means that only in the case of averaged equations over time one can consider the turbulent flow as partially two-dimensional. In reality, none of the components of flow velocity ever equals zero. However, we assume two-dimensionality for simplification.

5. Continuity: Change of the turbulence parameters is a continuous process. In turbulent flows, some processes increase and other processes weaken simultaneously.

6. High Reynolds number: A turbulent flow appears above a specific Reynolds number range.

From the previous discussion, we now know that TBL gives rise to the skin friction comparatively higher than the laminar ones. This is due to the fact that at sufficiently high Reynolds number the boundary layer flow becomes turbulent shortly after the leading edge. Therefore, large part of the wall encounters turbulence and its associated skin friction drag. Most of the subsonic passenger airliners are operated at the high Reynolds number, where most of the aerodynamic surfaces of fuselage, wings and different control surfaces are subjected to high Reynolds number. Figure-1.2 shows the operating range of the subsonic passenger airliners where vertical and lateral axis presents the operating velocity and their corresponding chord Reynolds number ($Re_C$) respectively. Here, $Re_C \geq 10^6$ is the regime where most of the general aviation operations are performed.

## 1.3 Turbulent Boundary Layer

Under certain conditions, such as at large Reynolds numbers laminar flow becomes unstable and eventually, turn into turbulent. To explain this phenomenon, one can theoretically explain as follows. The small perturbations overlap at the beginning of the laminar flow. As these perturbations grow with the Reynolds number, eventually the boundary layer changes into an unstable one due to increasing perturbation and finally, transforms into the turbulent form.

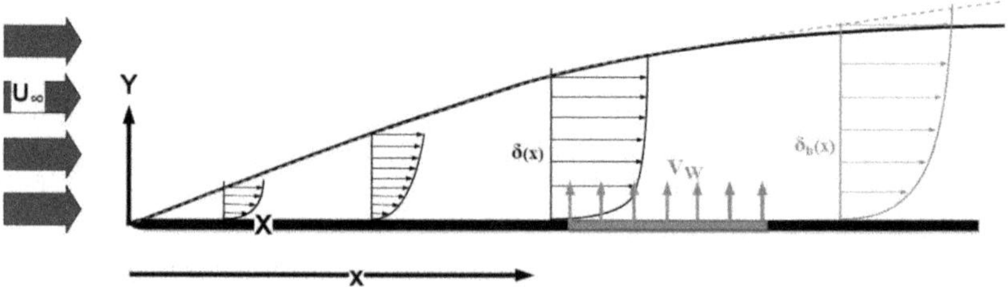


**Figure 1.3:** turbulent boundary layer developing on a flat plate with wall-normal transpiration from localized perforated surface (not scaled).

In order to distinguish between the classical turbulent boundary layer over non permeable surface and blowing boundary layer with perforated wall, reference turbulent boundary layer over a smooth wall will be referred as $_{SBL}$. A schematic of boundary layer developed over a flat plate is presented in Figure-1.3, where a perforated surface replaces the non permeable one in the red region where velocity magnitude of wall normal blowing is indicated with $V_W$.

The origin of the Cartesian co-ordinate system for presented one started at the leading edge of the flat plate, however, the ideal model of a turbulent boundary layer develops over a flat plate that is infinitely long along streamwise and spanwise direction. The ideal flat plate geometry which we refer to fulfills the zero pressure gradient condition, therefore, free stream velocity outside the boundary layer is given by $U_\infty$.

Here, boundary layer thickness (BLT) over Standard Boundary Layer (SBL) and blowing boundary layer is indicated with a solid black line and dashed red line respectively. They are termed as $\delta(x)$ and $\delta_b(x)$ respectively as a function of streamwise distance from the leading edge. Previous literature survey suggested that the Micro-blowing Technique gives rise to boundary layer thickness depending on the magnitude of the blowing velocity. Figure-1.3 exhibit the growth of the boundary layer (not scaled) with the dashed line when blowing is applied. Based on the hypothesis that boundary layer thickness is gradually increased due to blowing therefore, boundary layer thickness ($\delta_b$) is larger than the ones from Standard Boundary Layer e.g $\delta(x) < \delta_b(x)$.

## 1.4 Problem statement

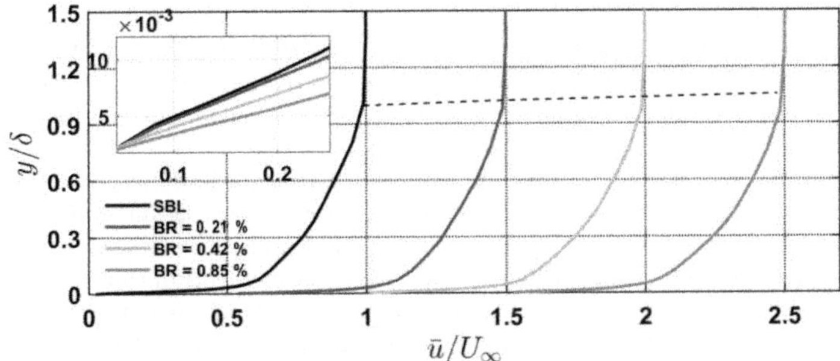


**Figure 1.4:** Profiles of a streamwise velocity profile scaled with the outer scaling parameter with different BR including reference Standard Boundary Layer at $Re_{\theta,SBL} = 1870$. Profiles are offset from each other along x axis by 0.5. Black "dashed" line indicate the points connecting the edge of boundary layer where mean streamwise velocity reaches 99% of the free stream velocity. Inset figure magnifies the linear slope in viscous sub-layer.

Blowing velocity ($V_W$) in wall normal direction is applied through the perforated surface from the perforated region as indicated with red color in Figure-1.3. $V_W$ was varied in different magnitudes in a range $0 < V_W < 6\%$ of the free stream velocity for different experimental cases. This is expressed as BR, is a non dimensional parameter defined as the ratio between the blowing magnitude and the free stream velocity while expressed in percentage.

$V_W$ is uniform in space, however, perforated surfaces must be used to approximate the ideal fully permeable surface for experimental realization. Therefore, by the term uniform blowing indicate uniformity of the blowing velocity in terms of the local spatial average. $V_W$ is constant in space at least over the center of the holes, in addition, no intentional variation in space are present other than the ones that is accompanied by the use of a perforated area. Although this can not be strictly confirmed for the surface between the holes where blowing velocity is zero.

Figure-1.4 presents the outer scaled profiles of streamwise velocity along wall distance. The profiles were measured at the same streamwise location with different BRs where BRs were gradually doubled. Although other boundary conditions were same, the "dashed" line connecting the edge of the boundary visibly shows gradual increase of BLT with increasing BR, which supports the assumption stated in the previous section. The slope of the mean velocity near the wall is also presented in the inset figure, it also clearly observable that the wall shear is also decreased with increased BR.

Unlike canonical pipe flow, BLT of a turbulent boundary layer is ever increasing. Directly after the transition to turbulent regime, the growth of BLT is larger compared to the laminar one. One way to understand the mechanism of boundary layer growth is to consider the exchange of momentum. The exchange of fluid momentum in the vertical direction away from the plate is greater in turbulent boundary layer than that of the Laminar Boundary Layer. This is because the loss of momentum at the wall which is diffused either by viscosity (e.g. molecular mixing) or by turbulent mixing. An analytical solution of the turbulent boundary layer equations of motions are not possible unless otherwise numerically. It is rather more suitable to use the semi-empirical techniques in experiments.

Although drag reduction is however, of primary interest, for flow control experiments in wall bounded shear flows, many of the flow properties such as integral properties, mean and turbulence statistics, are also changed. Therefore, in addition to the BLT, other integral properties such as displacement thickness, momentum thickness and wall shear are changed as well. Beside integral properties, blowing also affects turbulence properties of the boundary layer. Statistics, particularly mean and Root-Mean-Square of all the velocity profiles are strongly affected and to be specific, blowing enhances them. However, these issues related to the effect of blowing on the turbulent boundary layer, we can summarize some of those as following research questions.

- How far in downstream direction does blowing influence the mean flow characteristics?
- Is coherence effected in the outer region with increasing blowing ratios?

- Is the change of the outer peak in RMS velocity profile a fundamental change in turbulence characteristics due to blowing? Do we observe changes to the existing coherent structures, particularly structures in the outer region, or do we observe complete new types?
- Turbulent structures in the outer layer influence the inner region structures, to understand how blowing affects this interaction is an important goal.
- What are the optimal parameters of blowing that will result in the formation of a second peak in power spectra and cross spectra in the outer region and how to interpret this second peak?
- How to determine the optimal geometric arrangement for perforated surface in order to achieve maximum influence on TBL control with minimum energy expense? Whether distribution of perforation and streamwise extent of blowing region effect the mean properties of the flow?
- How TKE and its fractional distribution is effected along the wall with different blowing ratio?
- Do we see different energy levels in spectrogram with different blowing ratios? For structures bigger than 2-3 $\delta$, how does the distribution of spectral energy behave?
- Can we understand the physics of the reduction of wall shear stress and learn how to reduce friction drag by purpose of blowing application?

Turbulent boundary layer flows are of complex combination of different factors where the turbulent fluctuations are three dimensional which undergoes diffusion and transport of momentum. Energy of the turbulent structures present, have a continuous spectrum which are of self sustaining. There is a significant relationship between the production of Reynolds stress and these so-called turbulent structures. But blowing plays a strong role enhancing these Reynolds stresses and as a consequence, we expect significant modification to the frequency and energy aspect of these turbulent structures. In order to confirm such hypothesis, one has to look into the energy levels in spectrogram with different blowing ratios. Subsequently, how does the distribution of spectral energy behave.

Within the context of present thesis, we will deal primarily regarding the problems in canonical turbulent boundary layer associated with blowing such as scaling of blowing induced boundary layer, changes in their mean properties and turbulence statistics of the velocity profiles. One part of the present thesis deals with the inner scaling of the boundary layer profiles, friction parameters (wall shear velocity and friction coefficient) at moderate Reynolds number where as another part of the thesis discusses the turbulence statistics, their scaling, momentum transfer and Turbulent Kinetic Energy at high Reynolds number. Second part of this thesis also seeks to interpret the outer peak changes due to blowing. Blowing adds momentum flux into the flow, modify the diffusion process and enhances the Turbulent Kinetic Energy as a consequence.

Therefore, the region is investigated which is affected the most. In addition, we looked into the morphology of the coherent structures and investigated the effect of blowing.

As just said above, a TBL flow is very complicated and depends on many parameters and boundary conditions. The topic of such flow has been and still being investigated for centuries, many books are devoted wholly or in part to the theme of the turbulent flow. In a special series are the works of Rotta (1953), Tennekes (1972), Schlichting (1960), Pope (2000) and White (1974) where interested readers can find elaborate knowledge on the following topic.

### 1.4.1 Physics of a blowing induced Turbulent Boundary Layer

The boundary layer flow being investigated within the context of present thesis is assumed to be turbulent, incompressible e.g with constant viscosity and without heat transfer.

The continuity equation and simplified Navier-Stokes solution for incompressible, two-dimensional equation of motion representing the conservation of momentum and total shear stress for TBL under constant pressure is given below, where, the velocity components $\bar{u}(x, y)$, $\bar{v}(x, y)$ and $P_\infty$ represents the time averaged streamwise and wall-normal components of the velocity, the pressure outside the boundary layer respectively. $\mu$ and $\rho$ indicate the dynamic viscosity and density of air respectively. $u^{'}(t, x, y)$, $v^{'}(t, x, y)$ and $p^{'}(t, x, y)$ are the fluctuating components.

$$\frac{\partial \bar{u}}{\partial x} + \frac{\partial \bar{v}}{\partial y} = 0 \qquad \text{(continuity)}$$

$$\bar{u}\frac{\partial \bar{u}}{\partial x} + \bar{v}\frac{\partial \bar{u}}{\partial y} = -\frac{1}{\rho}\frac{\partial P_\infty}{\partial x} + \frac{\mu}{\rho}\frac{\partial^2 \bar{u}}{\partial y^2} - \frac{\partial \overline{u'v'}}{\partial y}\frac{\partial}{\partial x}(\overline{u'^2} - \overline{v'^2}) \qquad \text{(Navier-Stokes)}$$

$$(1.1)$$

Reynolds decomposition of the corresponding velocity component are:

$$u = \bar{u} + u^{'}, \qquad v = \bar{v} + v^{'} \tag{1.2}$$

Present thesis performed experiments on turbulent boundary layer over smooth surface in order to outline the reference characteristics. Therefore, boundary conditions at wall for reference turbulent boundary layer over an impermeable surface are defined with Equation-1.3.

$$\begin{aligned} &\bar{u} = 0 \qquad && \bar{v} = 0 \qquad && y = 0 \\ &\bar{u} = U_\infty && y = y && \end{aligned} \tag{1.3}$$

As we are currently dealing with the zero pressure gradient, therefore, the first term in the R.H.S of Equation-1.1(Navier-Stokes) e.g $\partial P_\infty / \partial x = 0$. The fourth term in the same equation is of secondary importance and will be neglected thereafter. This term becomes important as the flow reaches the point of separation, however, BRs used for the present experiment were selected as such that the flow never reached the point of separation. The second and the third component of the equation provides the wall normal variation of the total shear stress $\tau$ and is presented with the Equation-1.4. Integrating Equation-1.4 following this boundary condition, we obtain the expression for the total shear stress through Equation-1.5, where first and last term in the R.H.S present the *Viscous Shear Stress* (VSS) and *Reynolds Shear Stress* (RSS) respectively.

$$\frac{1}{\rho}\frac{\partial \tau}{\partial y} = \frac{\mu}{\rho}\frac{\partial \bar{u}}{\partial y} - \frac{\partial \overline{u'v'}}{\partial y} \tag{1.4}$$

$$\tau = \mu\frac{\partial \bar{u}}{\partial y} - \rho\overline{u'v'} \tag{1.5}$$

Further experiments were conducted over perforated surface with uniform blowing. In principle, boundary condition defined by Equation-1.3 has a modification when wall normal blowing at uniform rate is applied. Therefore, we can re-write Equation-1.3 as 1.6 which is the boundary condition under the influence of uniform blowing.

$$\begin{array}{lll} \bar{u} = 0 & \bar{v} = V_W & y = 0 \\ \bar{u} = U_\infty & y = y & \end{array} \tag{1.6}$$

The influence of fluid viscosity creates the wall shear stress ($\tau_w$) which extracts energy from the mean flow. The intermittent boundary between the potential flow and the boundary layer supply the energy required to feed into this wall shear stress.

$$\delta(x) = y(x, y), \bar{u} = 0.99U_\infty \tag{1.7}$$

This takes place at the edge of the boundary layer. A general description of this edge separating the non rotating potential flow is relevant which is described with the help of the BLT. This is presented in Equation-1.7.

It is known as the distance from the wall where the time averaged mean streamwise velocity reaches its 99%. Although, BLT of the turbulent boundary layer is presented with a mean value, in reality this is strongly intermittent in space and unsteady in time. Therefore, requires higher spatial resolutions with smaller velocity difference and subsequently, interpolation of the data points of the boundary layer velocity profile. Hence, uncertainty estimation is high for the statistical representation of the data. Figure-1.5 presents the BLT detection method visually using data obtained at $Re_\theta = 1870$.

It is often preferred to describe the turbulent boundary layer with the help of Integral properties of the turbulent boundary layer such as *displacement thickness* and *momentum thickness*. Such integral properties offer less error compared to the ones obtained from the interpolated values such as BLT. Pohlhausen (1921) derived a simplified method to calculate these integral parameters by solving Equation-1.1. The solution is also known popularly as momentum integral equation. Equation-1.8 presents the displacement thickness ($\delta^*$).

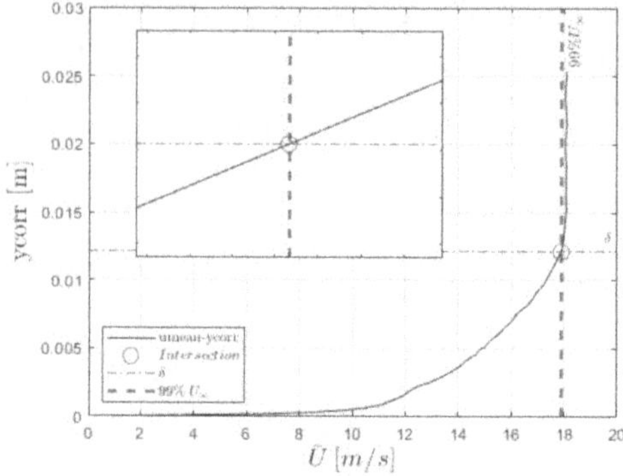


**Figure 1.5:** Dimensional streamwise velocity distribution along the wall distance using LDA method. Inset figure magnifies the region of interpolation with the $\bar{u} = 0.99U_{\infty}$.

Due to the frictional dissipation, the boundary layer can be considered to possess a total momentum flux deficit. This momentum loss in comparison to the potential flow is expressed as momentum thickness ($\theta$) through Equation-1.9. The upper limit of the wall distance in Equation-1.8 and 1.9 is suggested to be set at the BLT. For determination of the integral length scales within the present thesis, we have set the outer limit of integration as the wall distance where mean streamwise velocity reaches 99%.

Rotta (1950) proposed a different length scale using the free stream velocity, wall shear velocity and the displacement thickness knows as Rotta-Clauser length scale ($\Delta$) as expressed in Equation-1.10. The Rotta-Clauser length scale is also derived using an integral length scale e.g displacement thickness and therefore, easier to calculate from experimental data with relatively few number of data points along the boundary layer. Here, $u_\tau = \sqrt{(\tau_w(x)/\rho)}$. $\tau_W(x) = \mu(d\bar{u}/dy)$ is the local shear stress at wall which depends on the distance from the leading edge of the plate. Subsequently, we derive the skin friction coefficient by normalizing wall shear stress with the dynamic pressure of the free stream e.g $C_f = \tau_W/0.5\rho U_\infty^2$. In order to verify the experimental results, Smits et al. (1983) proposed an empirically derived power law for turbulent boundary layer flows over smooth wall and expressed as $C_f Re_{\theta,SBL} = K$, where constant $K = 0.024$.

$$\delta^* = \int_{y=0}^{\infty} (1 - \frac{\bar{u}}{U_\infty}) dy \tag{1.8}$$

$$\theta = \int_{y=0}^{\infty} \frac{\bar{u}}{U_\infty} (1 - \frac{\bar{u}}{U_\infty}) dy \tag{1.9}$$

$$\Delta = \frac{U_\infty \delta^*}{u_\tau} \tag{1.10}$$

In order to derive an expression of the skin friction co-efficient, we recall the boundary layer momentum equation (Equation-1.1). Von Kármán obtained an expression of skin friction co-efficient (derived from Equation-1.1) and presented here as Equation-1.11. In presence of wall normal transpiration, Equation-1.11 present the most compact form of the momentum integral equation for 2D incompressible flow. Since, free stream velocity does not very along the streamwise direction for zero pressure gradient condition and for smooth wall (e.g $dU_\infty/dx = 0$ and $V_W = 0$), therefore, second and third term in Equation-1.11 is zero. Finally, we obtain Equation- 1.12 for zero pressure gradient turbulent boundary layer over smooth wall.

$$\frac{C_f}{2} = \frac{d\theta}{dx} - \frac{V_W}{U_\infty} + (2+H)\frac{\theta}{U_\infty}\frac{dU_\infty}{dx} \tag{1.11}$$

$$= \frac{d\theta}{dx} - \cancelto{0}{\frac{V_W}{U_\infty}} + (2+H)\frac{\theta}{U_\infty}\cancelto{0}{\frac{dU_\infty}{dx}}$$

$$C_f = \frac{2d\theta}{dx} = \frac{2\tau_W}{\rho U_\infty^2} = 2\left(\frac{u_\tau}{U_\infty}\right)^2 \tag{1.12}$$

Detailed determination process for wall shear stress, wall shear velocity and friction co-efficient is discussed in the section that follows.

$$H = \frac{1}{1 - C'/U_\infty^+} \tag{1.13}$$

where,

$$C' = \int_{y=0}^{\infty}\left(\frac{U_\infty - \bar{u}}{u_\tau}\right) d\left(\frac{y}{\Delta}\right)$$

and, $$U_\infty^+ = \frac{U_\infty}{u_\tau}$$

The turbulent boundary layer property $H$ in Equation-1.11, which is the ratio of both displacement thickness and momentum thickness is known as the shape factor. This is the rough indication of the shape of the boundary layer velocity profiles and indicate whether the boundary layer is turbulent, although it is not always a definitive parameter for exact identification of turbulent boundary layers from the laminar ones. A large shape factor indicate a boundary layer approaching separation. Schlichting (1960) set the value of the shape factor as $H = 2.59$ for the laminar region which decreases to $H \approx 1.4$ in the turbulent region. For verification of the experimental data, Nagib et al. (2007) proposed an empirical relationship based on Equation-1.13.

Finally, various Reynolds number used within this thesis are defined using different length scales such as streamwise distance, displacement thickness, momentum thickness and wall shear velocity and BLT.

In order to describe the flow condition in wall bounded external flows such as boundary layer flow, the non dimensional parameter Reynolds number is expressed to describe the outer flow condition based on the characteristics length (X) and the free stream velocity. This is the only parameter that is independent of BR, but local inertial and momentum effect is rather comprehensive in terms of shear Reynolds number ($Re_\tau$ or $\delta^+$) and momentum thickness Reynolds number ($Re_\theta$) respectively.

$$Re_x = \frac{U_\infty x}{\nu}, \quad Re_{\delta^*} = \Delta^+ = \frac{U_\infty \delta^* / u_\tau}{\nu / u_\tau} = \frac{U_\infty \delta^*}{\nu}, \quad Re_\theta = \frac{U_\infty \theta}{\nu}, \quad Re_\tau = \delta^+ = \frac{u_\tau \delta}{\nu}, \tag{1.14}$$

Previous researches and preliminary results presented here indicate significant changes in local wall shear and time averaged mean momentum under stochastic influence of the perturbation due to Micro-blowing. Hence, in the subsequent discussion both of these parameters will be taken into consideration to explain previous research effort.

### 1.4.2 Prandlt's mixing length hypothesis

Prandlt (Prandlt (1927)) postulated that at the wall region molecules of fluid striking the wall give up their kinetic energy e.g. the wall absorb the kinetic energy of the turbulent fluctuations of the velocity vectors. Consequently, near wall smaller eddies are expected to die. Prandlt also hypothesized that the mixing length (or the eddy size) is proportional to distance from wall in the logarithmic region. Average mixing length is reduced as the solid wall is approached. Simplest relationship in order to quantify this phenomenon is expressed with Equation-1.15, where, $l_m$ is the mixing length and $\kappa$ is the proportional constant also known as mixing length constant. Von Kármán determined this constant empirically, therefore, it is also known as Kármán constant.

$$l_m = \kappa y \tag{1.15}$$

Prandlt (1927) postulated that total shear stress ($\tau$) at any given point within the boundary layer is given by the sum of the molecular and turbulent contribution (VSS and RSS), therefore we come back to Equation-1.5 which presented us the total shear stress can be written as Equation-1.16.

$$\tau = \underbrace{\mu \frac{d\bar{u}}{dy}}_{\text{molecular component}} + \underbrace{\rho l_m^2 \left| \frac{d\bar{u}}{dy} \right| \frac{d\bar{u}}{dy}}_{\text{turbulent component}} \quad (\text{where, } \mu = \rho\nu) \tag{1.16}$$

The effect of the molecular transport is small compared to that of the turbulent one and neglected in comparison to the turbulent component. Therefore, shear stress is assumed constant and equal to the second term in the Equation-1.16.

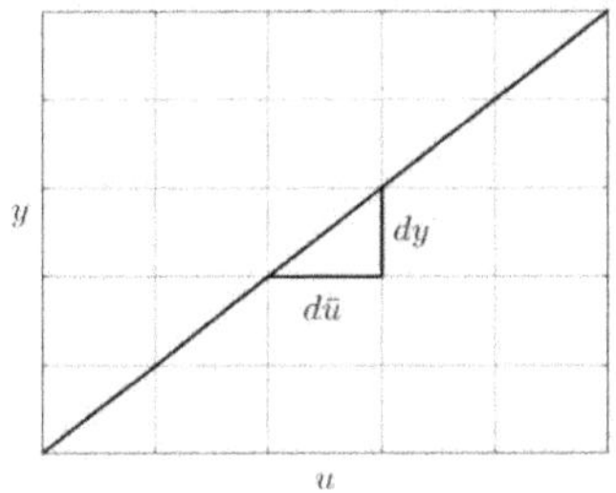


Figure 1.6: Slope in the linear sub-layer

$$\tau = \tau_w = \rho l_m^2 \left(\frac{d\bar{u}}{dy}\right)^2$$
$$= \rho \left[ l_m \left(\frac{d\bar{u}}{dy}\right)\right]^2$$

Therefore,

$$\sqrt{\left(\frac{\tau_w}{\rho}\right)} = l_m \left(\frac{d\bar{u}}{dy}\right) = u_\tau \quad (1.17)$$

We derive the wall shear velocity through Equation-1.17 or the local friction velocity ($u_\tau$) based on the shear stress at wall ($\tau_w$).

In order to obtain law of wall relationship at the logarithmic region, using Equation-1.15 and 1.17, we can replace the value of $l_m$ and obtain the expression for the law of wall through Equation-1.18.

$$\kappa y \left(\frac{d\bar{u}}{dy}\right) = u_\tau$$
$$\frac{d\bar{u}}{u_\tau} = \frac{dy}{\kappa y}$$
$$\frac{1}{u_\tau}\int d\bar{u} = \frac{1}{\kappa}\int \frac{dy}{y} \qquad \text{[Integrating]}$$
$$\frac{\bar{u}}{u_\tau} = \frac{1}{\kappa} ln(y) + C \quad (1.18)$$

This relationship generally states that the streamwise velocity is proportional to the log of wall distance, except for the viscous sub-layer near the wall. Here, $C$ is the integration constant. One has to determine the value of $\kappa$ and $C$ experimentally. However, commonly accepted value are 0.4 and 5 respectively. For present thesis value of $C$ and $\kappa$ differed by 4.127 and 0.384 respectively. This value is valid for all the measurements obtained using LDA. $\kappa$ and $C$ was determined by fitting $\Xi$ and $\psi$ to the wall distance using the Equation-1.19 for present experiment using LDA. This will be discussed in the results section.

$$\Xi = \left(y^+ \frac{d\bar{u}^+}{dy^+}\right) \qquad \text{(inner scaled)}$$
$$= y\left\{\frac{d(\bar{u}/U_\infty)}{dy}\right\} \qquad \text{(outer scaled)}$$
$$\psi = \bar{u}^+ - \frac{1}{\kappa} lny^+$$

From Equation-1.18, when $y \to 0$, $ln(y) \to \infty$, at $y = y_0$, $u = 0$, replacing the values in Equation-1.18 we obtain,

$$C = -\frac{1}{\kappa}ln(y_0) \tag{1.19}$$

From Equation-1.18 and 1.19,

$$\begin{aligned}\frac{\bar{u}}{u_\tau} &= \frac{1}{\kappa}ln(y) - \frac{1}{\kappa}ln(y_0)\\ &= \frac{1}{\kappa}\{ln(y) - ln(y_0)\}\\ &= \frac{1}{\kappa}ln\left(\frac{y}{y_0}\right)\end{aligned} \tag{1.20}$$

Here, $y_0$ is regarded as the shift in the hydraulic wall and is a function of the surface roughness condition ($y_0 \propto \nu/u_\tau$). This describes the velocity profile in the outer region only.
Stated earlier,

$$\begin{aligned}y_0 &\propto \frac{\nu}{u_\tau}\\ y_0 &= \beta\frac{\nu}{u_\tau}\end{aligned} \tag{1.21}$$

From Equation-1.20 and 1.21, hence we obtain the law of the wall:

$$\begin{aligned}\frac{\bar{u}}{u_\tau} &= \frac{1}{\kappa}ln\left(\frac{y}{\beta\frac{\nu}{u_\tau}}\right)\\ &= \frac{1}{\kappa}ln\left(\frac{yu_\tau}{\beta\nu}\right)\\ \text{or, } \frac{\bar{u}}{u_\tau} &= \frac{1}{\kappa}\left[ln\left(\frac{yu_\tau}{\nu}\right) - ln\beta\right]\end{aligned} \tag{1.22}$$

Equation-1.22 is the universal, dimensionless, logarithmic velocity distribution law as Prandlt postulated. Experimentally, $\beta = 0.111$. This equation is valid where viscous shear is neglected (log-layer).

Gradient of the mean streamwise velocity at the near wall region is an important parameter to determine both *Viscous Shear Stress* and *Turbulent Kinetic Energy*. However, $d\bar{u}/dy$ depends on $y^+$ and $y/\delta$. Based on the dimensional analysis from the book of Pope (2000), we can obtain Equation-1.23. Where, $\Phi$ is an universal non dimensional function.

As we reach far away from the wall e.g $y^+ \gtrsim 50$, where outer length scale becomes important to describe the flow. For sufficiently large Reynolds number (for example $Re_x \geq 20 \times 10^3$ or $Re_\theta \geq 1700$ (Mathis et al. (2009))), R.H.S of Equation-1.23 is independent of $y^+$. Replacing the R.H.S value in Equation-1.23 and integrating between the wall boundary for $y$ and $\delta$, we obtain the traditional form of "*Velocity defect law*" as

Equation-1.24. "Velocity defect" by definition is the difference between the mean streamwise velocity and the free stream velocity. In other words, velocity defect normalized by wall shear velocity depends only on $\eta = y/\delta$.

Millikan (1938) proposed the velocity profiles in the law of the wall by matching the *velocity defect law.* A logarithmic velocity distribution results in the overlap region ($\delta \gg y \gg \nu/u_\tau$) at sufficiently high Reynolds number as discussed above. A logarithmic velocity profile is given following Equation-1.25 for a smooth wall turbulent boundary layer. Where, $B_1$ is the velocity-defect constant.

turbulent boundary layer flows which follows Equation-1.24 without showing Reynolds number dependency in the outer region of the boundary layer are also known as *equilibrium / self preserving boundary layer.*

$$\frac{d\bar{u}}{dy} = \frac{u_\tau}{y}\Phi\left(y^+, \frac{y}{\delta},\right) \tag{1.23}$$

$$\lim_{y^+ \to 0} \Phi\left(y^+, \frac{y}{\delta}\right) = \Phi_0\left(\frac{y}{\delta}\right)$$

$$= \frac{u_\tau}{y}\Phi_1\left(\frac{y}{\delta}\right)$$

$$\frac{U_\infty - \bar{u}}{u_\tau} = \Phi_1\left(\frac{y}{\delta}\right) \tag{1.24}$$

$$= -\frac{1}{\kappa} ln\left(\frac{y}{\delta}\right) + B_1 \tag{1.25}$$

Let us come back to Equation-1.18, which defines a universal law for the logarithmic region. When we plot the streamwise velocity velocity profiles along the wall distance for various data sets, this law defines a overlapping region (also known as logarithmic region) which only depends on the $\kappa$ and $C$. Coles (1956) formulated the "law of the wake" for the region in the TBL where the velocity profiles reaches further away from this logarithmic region. Based on Equation-1.18, he proposed the law of the wake as a function ($\mathbb{W}$). This is presented as follows:

$$\frac{\bar{u}}{u_\tau} = f\left(\frac{yu_\tau}{\nu}\right) + \frac{\Pi}{\kappa}\mathbb{W}\left(\frac{y}{\delta}\right) \tag{1.26}$$

$\mathbb{W}$ is continuously varying profile that is linear near the wall and logarithmic as it goes away from the wall. In the R.H.S of the Equation-1.26, $\Pi$ represent the wake parameter (actually is a function of pressure gradient) and can be obtained from Equation-1.27 using Equation-1.18 and 1.26. There is also an alternative to obtain $\Pi$ from Equation-1.25 e.g $B_1 = 2\Pi/\kappa$. Typically, for zero pressure gradient (ZPG) TBL $\Pi = 0.45$. At $y = \delta$, mean streamwise velocity reaches $\bar{u} = U_\infty$, therefore, replacing the boundary condition we obtain:

$$\frac{U_\infty}{u_\tau} = \frac{1}{\kappa} ln\left(\frac{yu_\tau}{\nu}\right) + C + \frac{2\Pi}{\kappa} f\left(\frac{y}{\delta}\right) \tag{1.27}$$

In order to describe the mixing length hypothesis at wall, as stated in the first part of the Equation-1.16, we can re-write:

$$\begin{aligned}
\tau = \tau_w &= \mu\frac{\bar{u}}{y} && \text{( by definition, } u_\tau = \sqrt{\tau_w/\rho} \text{ and therefore, } \tau_w = u_\tau^2\rho) \\
u_\tau^2\rho &= \mu\frac{\bar{u}}{y} \\
\frac{\bar{u}}{u_\tau} &= \frac{yu_\tau}{\nu} && \text{(re-arranging and we know, } \nu = \mu/\rho) \\
\bar{u}^+ &= y^+
\end{aligned} \tag{1.28}$$

This Equation-1.28 holds valid for the viscous sub-layer e.g the region near the wall, viscosity begins to dominate the shear stress, and the velocity is proportional to wall distance.

## 1.4.3 Van Driest profile

Van Driest (1956) proposed an analytical relationship from the equation of total shear stress based on the empirical data from TBL in order to represent the law of wall. The analytical relationship is expressed using Equation-1.30, where, $A^+ = 26$, The value of Von Karman constant ($\kappa$) is taken as 0.40. This equation is particularly important in order to verify the experimental data obtained in TBL over smooth surface. $l_m^+$ represent the non-dimensionalized mixing length. This relationship is particularly applicable to the buffer region immediately after the viscous sub-layer.

$$\bar{u}^+(y^+) = \int_0^{y^+} \frac{2dy^+}{1 + \sqrt{1 + 4\kappa^2 y^{+2}[1 - exp(-y^+/A^+)]^2}} \quad \text{for } 0 \le y^+ \le 55 \tag{1.29}$$

$$l_m^+ = \kappa y^+[1 - exp(-y^+/A^+)] \tag{1.30}$$

### 1.4.4 Summary of law of the wall description

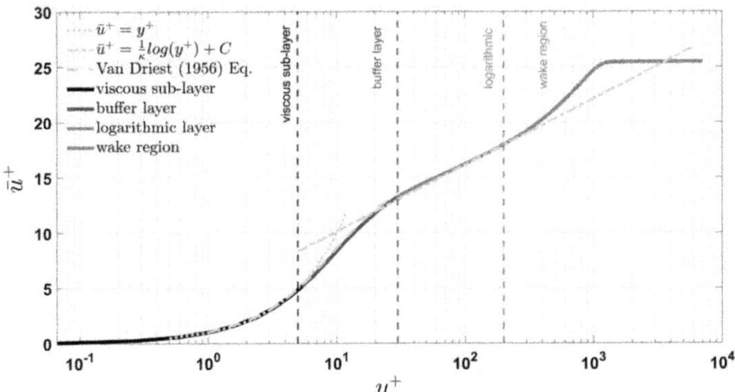


**Figure 1.7:** Typical TBL profile using inner length scales where viscous sub-layer and logarithmic layer is presented with Equation-1.28 and Equation-1.18 respectively.

Finally, mixing length hypothesis can explain mechanisms by which transport takes place in the TBL. Based on the hypothesis one can apprehend a picture of multi-layer TBL as shown in Figure-1.7 which is divided into four regions. Wei et al. (2005) suggested the extent of these layers and are as follows:

- A "viscous sub-layer" with a linear velocity profile in the immediate vicinity of the solid boundary e.g $y^+ < 5$. Here, molecular transport is the dominant one and the streamwise velocity can be expressed with Equation-1.28.
- A "buffer layer", which extent upto $5 < y^+ < 30$. Here the contributions of the molecular and turbulent transport are of comparable magnitude. Follows Equation-1.30 from Van Driest (1956).
- A turbulent outer region. This is comprised of a logarithmic and a wake region. In order to depict this region clearly Equation-1.25 or *velocity defect law* presentation is advised.
- The turbulent "logarithmic layer" as stated in the Equation-1.18, this can be detected as $30 < y^+ < 0.15\delta^+$.
- A "wake region" where $y^+ > 0.15\delta^+$ which follows the law of the wake (Equation-1.27).

- The velocity profiles in each layer were found to be dependent on the friction velocity ($u_\tau$), which implied that determination of the wall shear stress is a priori.

In addition, Alfredsson and Örlü (2010) and Alfredsson et al. (2011) proposed a new scaling of boundary layer profile data in order to avoid uncertainties that arises from the sensitive wall shear stress measurements. This scaling is based on the Equation-1.31 with the wake region ($\bar{u}/U_\infty \geq 0.71$). Alfredsson and Örlü (2010) proposed the values of constants in Equation-1.31 as a=0.031 and b=0.260, whereas fitting of present SBL data derives a=0.01 and b=0.240.

$$\frac{u_{rms}}{U_\infty} = a + b\left(1 - \frac{\bar{u}}{U_\infty}\right) \tag{1.31}$$

$$\frac{u_{rms}}{\bar{u}} = c + d\left(1 - \frac{\bar{u}}{U_\infty}\right) \tag{1.32}$$

### 1.4.5 Sutherlands correction

In order to obtain accurate estimation of the air viscosity data, particularly the kinematic viscosity of air ($\nu$), Sutherland (1893) proposed a relationship between the dynamic viscosity ($\mu$), and the absolute temperature ($T$) and density ($\rho$) of an ideal gas such as air. This relationship is also known as Sutherlands law. After obtaining the air properties such as temperature ($T$) in Kelvin, atmospheric pressure ($P_{atm}$) in Pascal and static pressure ($P_{st}$) in Pascal during experiment, this is applied to estimate the $\mu$ and $\rho$, using Equation-1.33 and Equation-1.34 respectively. Finally, replacing the value of $\mu$ and $\rho$ in Equation-1.35, we obtain the kinematic viscosity ($\nu$).

$$\mu = \frac{C_1 T^{3/2}}{T + S} \tag{1.33}$$

$$\rho = \frac{P_{atm} + P_{st}}{RT} \tag{1.34}$$

$$\nu = \mu/\rho \tag{1.35}$$

Where, in Equation-1.33, $S$ is the Sutherland temperature which is 110.4 $K$ and $C_1$ is the Sutherland constant which is $1.458 \times 10^{-06}$ $kg/m.s.\sqrt{K}$. In Equation-1.34, $R$ is the ideal gas constant and the value is 286.8.

### 1.4.6 High Reynolds number

From Figure-1.2, we can see that large subsonic jet aircraft's are operated within chord Reynolds number range up to and beyond $10^7$ (Lissaman (1983)). In order to improve the drag performance of aircraft surface, it is also necessary to evaluate the performance of the applied control technique in such a way that the performance parameters are comparable to the operating range. In contrast, experimental conditions are far beyond the operating range for control experiments in laboratory. Although, experiments provide important insight for the application but appropriate utilization of such results are still a matter of debate.

Fernholz and Finley (1996) presented the Normalized RMS Streamwise velocity fluctuation in different wall region from the HWA measurements obtained from incompressible, ZPGTBL at high Reynolds number. Data display that measurements beyond $Re_\theta > 7140$ have an outer peak. For TBL study, whether intended for the flow control or high Reynolds number dependency investigation, classical threshold for optimum shear Reynolds number > 2000 was suggested by Smits et al. (2011). A slightly different value of sufficient "High Reynolds number" estimate to observe a decade of separation between near wall and log-region coherent structures was proposed by Hutchins and Marusic (2007a). Based on the observations from HWA measurements at high Reynolds numbers, they have suggested that at $Re_\tau > 1700$ e.g at a critical value of $Re_\tau = 167$, inner and outer sites will have a sufficient separation. Therefore, the existence of a wider overlap layer and sufficient separation between the scales at highly turbulent regimes provide a clear and more distinguished observation of structures with different length scales. This challenges the traditional point of view that the inner layer cycle of turbulence is independent of the outer layer influence (Jiménez and Moin (1991), Hamilton et al. (1995), Waleffe (1997), Jiménez and Pinelli (1999)). Nevertheless in the recent years, an opposing argument regarding the autonomy of the inner region (viscous sublayer) where high and low speed streaks are the dominating turbulence structure has emerged. Recent scientific works suggests that the outer layer influences the inner layer and it keeps growing as the Reynolds number grow.

In order to elaborate the effect of outer layer on the inner one in wall bounded flows, strong shear layer near the wall causes extreme fluctuation. As a result, turbulent production process is substantially modulated from the wall roughness. But as we go beyond a certain threshold of the Reynolds number where inner and outer region has sufficient separation in terms of their peak value, outer layer influence on the energy scale become comparable to the inner layer. With increasing Reynolds number beyond the threshold value, outer layer cycle reigns over the inner layer influence. HWA measurements at high Reynolds number facility from Hutchins and Marusic (2007a) indicate that shear Reynolds number of the experiment should be $\mathrm{Re}_\tau = 1700$ in order to observe at least one decade of separation in the streamwise fluctuation data. Where, most energetic peak location both in turbulence intensity and energy spectra is than under the scope of measurement to study. On the other hand uncertainty regarding inner peak location increases at high Reynolds number due to the probe effect for HWA measurements (Hutchins et al. (2009)). Therefore, studying the outer region using non-intrusive technique such as PIV offers comparatively less uncertainty and easy handling of large volume of measurement data (Tang et al. (2019)).

One of the traditional focus of the wall-bounded turbulence studies had been and still, concentrated on scaling and analysis of turbulence intensity profiles which also represent diagonal components of Reynolds stress tensor. Experiments of Fernholz and Finley (1996) and Österlund et al. (1999a) under ZPGTBL have revealed the existence of an inner peak, where turbulent kinetic energy production reaches its maximum. The inner peak in boundary layer flow grows with increasing Reynolds number indicate a growing outer layer influence on the near-wall motions. It is now commonly accepted that the logarithmic layer plays an important role (Hutchins and Marusic (2007a) and Mathis et al. (2009)). Specifically at sufficiently high Reynolds numbers, it is the logarithmic region which contributes most to the bulk turbulent kinetic energy production as well as Reynolds stresses. It is therefore, necessitates a description of the turbulent structures that are present in the TBL.

## 1.5 Study of Coherent Structures in Turbulent Boundary Layers

"Big whirls have little whirls,
That feed on their velocity;
And little whirls have lesser whirls,
And so on to viscosity.
- Lewis Fry Richardson
"

Although there is no generally accepted definition for Coherent structures, in an effort to explain such structures, a definition which Cantwell (1981) proposed and rephrased by Adrian (2007), such as *"Complex, multi-scaled, random fields of turbulent motion down into more elementary organized motions that are variously called eddies or coherent structures".* This was also named as "Organized Structure". Theodorsen (1952) (Figure-1.8) depiction of horseshoe and Robinson's review (Robinson (1991)) are some of the primary depiction of these coherent structures.

Shear flow turbulence is dominated by a quasi-periodic sequence of large-scale structures often referred to as turbulent coherent structures. Coherent structures are not only quasi-periodic, but are different in size and shape depending on the location of these structures within the flow. Coherent structures are born, grow and die within the boundary layer, these evolves both in space and time. In a more recent study, Smits et al. (2011) classified these so called Coherent Structures in four characteristics elements, namely *near wall streaks* and *hairpin/horse shoe vortex* which is already discussed. Larger elements are termed as Large Scale Motion (LSM) and Very Large Scale Motions (VLSM).

When flow Reynolds number is high enough to enter into turbulent regime, hairpin vortices begins to form within a strong shear layer and create low and high-speed regions between them. Figure-1.8 (a) depicts different kind of structures found typically within the respective shear layer. The low speed regions which is close to the wall, termed streaks, grow downstream and develop inflectional velocity profiles. Simultaneously, the interface between the low and high-speed region begins to oscillate, promote the onset of a secondary instability. The low speed region starts to lifting up away from the wall when the oscillation amplitude keeps increasing and the flow starts to rapidly breaking down (these event is termed as 'bursts') to a motion which is completely chaotic. In the sequence of turbulent activities within the boundary layer, there are two important events, , called sweeps (or inrushes) and ejections (bursts). More elementary organized motions that are variously called eddies or coherent structures. The nomenclature of the basic turbulent structures found in different wall layers as originally introduced by Theodorsen is depicted in Figure-1.8 (a). Their interaction is visually depicted in the Figure-1.8 (b).

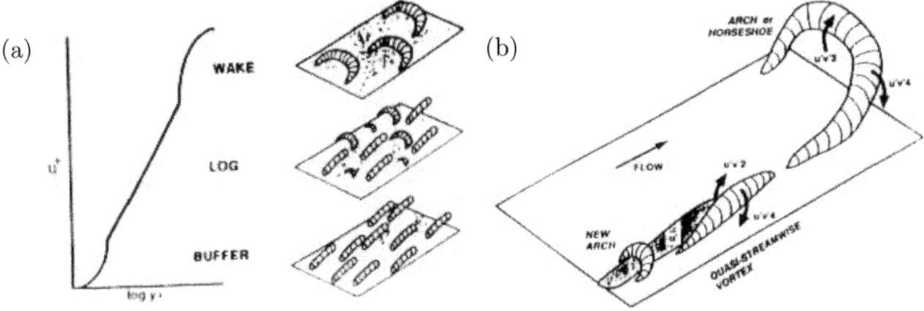


**Figure 1.8:** (a) Different shapes and sizes of coherent structures present in the TBL, (b) Mechanism of sweep and ejection events to these coherent structures (Theodorsen (1952))

Eddy hypothesis proposed by Townsend (1976), as hairpins form, they lift the quasi-streamwise vortices and create "lifted" hairpins that may appear to be "detached," depending upon the visualization method. The growth of the packets provides a mechanism to transport vorticity, low momentum,and turbulent kinetic energy from the wall. However, the transport cannot be exclusively due to the coherent structure described here, because turbulence is also produced by gradients away from the wall, and the later production may be due to a different mechanism.

A first experimental study of the near-wall turbulent structures were reported by Kline et al. (1967). A vivid understanding through visualizations of the *near wall streaks* which have a typical span-wise spacing in the order of viscous wall units approximately $100y^+$ was described. A further extension of the former was extended by Kovasznay et al. (1970) over the intermittent outer region and to explain the mechanism of the large eddies, hence, Hot Wire Anemometry (HWA) measurements confirms the existence of LSM's using conditional sampling and averaging. Later, Falco (1977), Head and Bandyopadhyay (1981) and Brown et al. (1977) extensively studied such motions using flow visualization and HWA respectively. In contrary, Brown approached with an individualistic view towards the mechanism of turbulent bulges or large structures and Falco identified the period of occurrence of this large scale structures with the outer scaling of 2.5 $\delta/U_\infty^2$. Head and Bandyopadhyay (1981) proposed a model of these so called large scale motions having a span-wise width of 100 $\nu/u_\tau$. Illustration of such mechanism of large structures was summarized by Adrian (2007) and presented here as Figure-1.9(left). He argued that the most prominent of such large structures are most common in the outer region especially in the logarithmic region and intermittently observed beyond the edge of the boundary layer.

In the study by Hutchins and Marusic (2007) where HWA measurements are conducted at high Reynolds numbers, this experimental study particularly challenges the traditional viewpoint that the inner cycle is independent of the outer cycle (e.g viscous layer is independent of the outer layer). Therefore, active modulation of the near wall scale is certainly an outcome of the outer region large-scale events. Nevertheless in the recent years, an opposing argument regarding the autonomy of the inner region (viscous sub-layer) where high and low speed streaks are the dominating turbulence structure has emerged Jiménez and Moin (1991), Hutchins and Marusic (2007), Hamilton et al. (1995), Waleffe (1997), Jiménez and Pinelli (1999) and Kim et al. (2011).

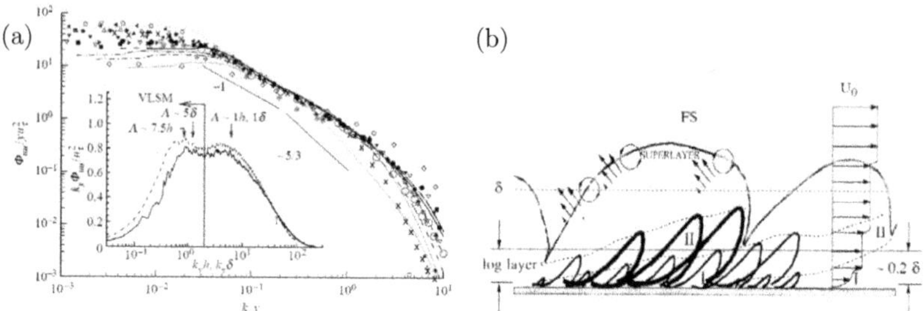


**Figure 1.9:** (a) Power spectra of streamwise velocity fluctuations, Small broken line: $Re_\tau = 1476, y/\delta = 0.05$ , solid line: $Re_\tau = 2395, y/\delta = 0.05$, solid line: $Re_\tau = 1476, y/\delta = 0.05$ (Balakumar et al. (2007)), (b) Summary sketch of the organization of hairpins and packets in a boundary layer (Adrian (2007)).

Spectral analysis of Balakumar et al. (2007) is presented as the power spectra in Figure-1.9 (a) and exhibit good collapse of data when reproduced with pre-multiplied power spectra. Therefore, this experimental study proved that Streamwise velocity fluctuation in wave number space is an effective method to identify large scale coherent motions.

Apart from the study related to energy production and dissipation data (probe data averaged in time), study of these so-called coherent structures are also found important in boundary layer dynamics Robinson (1991). The sketch of the these large motions are depicted in Figure-Figure-1.9 (b). The morphology of the structures had been extensively studied by Adrian (2007), Adrian (2000) and Adrian (2000), streamwise and wall-normal plane in boundary layer and channel flows had been largely studied and determination technique of hairpin packet (in log and wake region) was developed with lesser error mergin. The mechanism for the primary, secondary and tertiary hairpin like vortex packets in channel flow is presented in Figure-1.10, However, Direct Numerical

Simulation (DNS) is limited to low Reynolds number and hence, necessitates supportive experimental study. Individual eddy structures were experimentally investigated in TBL using Stereo Particle Image Velocimetry (SPIV) technique in ref Foucaut et al. (2014), Carlier and Stanislas (2005) and Herpin et al. (2013).

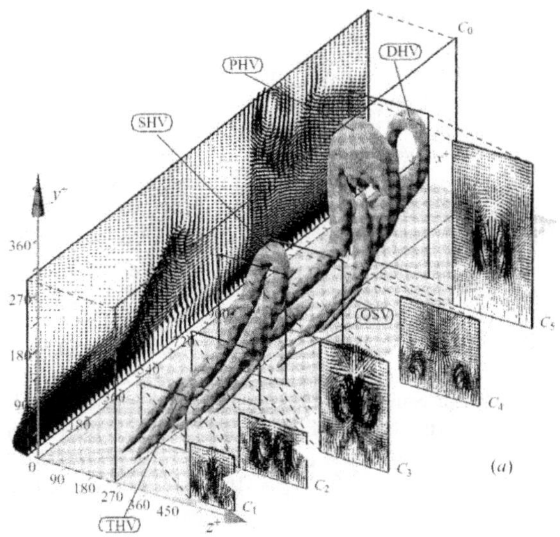


**Figure 1.10:** Hairpin like structure packets in channel flow at $Re_\tau$ = 180 (Zhou et al. (1999)).

A paramount progress has been made with the numerical study in TBL. Flow control and subsequent eddy vortex generation in spanwise-wall normal plane is presented in Spalart (1988). A list of reference numerical investigation can be found in the reference section (Kasagi and Shikazono (1995), Zhou et al. (1999), Wu and Moin (2009), Schlatter et al. (2009a), Schlatter et al. (2009b), Schlatter et al. (2010a), Schlatter et al. (2010b), Schlatter and Örlü (2013) and Eitel-Amor (2014)). Here, Schlatter and colleagues has contributed at pace in terms of higher Reynolds number DNS data. A large collection of data based on DNS upto $Re_\theta$ = 4060 (Schlatter et al. (2010a)) and LES upto $Re_\theta$ = 8300 (Eitel-Amor (2014)). The quality of the data compared to the experiment is in good agreement in terms of statistics. Study by Zhou et al. (1999) using DNS in TBL shed light on the packets of hairpin like structures as presented in Figure-1.10. Here, visualization of the packets were done using the imaginary part of the complex eigenvalue of the velocity gradient tensor. In addition, vortex core from the cross sectional view along YZ plane is also magnified.

One of the major aspect to deal with wall bounded shear flows (e.g. Boundary layer over flat plate, channel and pipe configuration) research is to study the dynamics of turbulent structures by the means of statistical analysis. In recent years, outer layer study has become more prominent in terms of their increasing contribution to the overall energy content. Such outer layer is mostly populated with Large and very large scale structures (LSM and VLSM) that can be extended upto twice/thrice of the boundary layer thickness ($\delta$). Despite exhaustive research effort, our understanding of such phenomena considering their physical mechanism is quite limited to develop models for effective flow control.

When considering a TBL over smooth and solid surface in order to realize the importance of these structures one needs to look into the turbulence statistics of the measurement data. Therefore, Fernholz and Finley (1996) measured the turbulence statistics in a wide range of Reynolds number using two different wind tunnel facilities (German-Dutch wind tunnel (DNW) and TU Berlin (HFI)). Normalized RMS of Streamwise velocity fluctuation in different wall location is given at Figure-1.11 adapted from Fernholz and Finley (1996) (here, $Re_{\delta_2}$ indicates momentum thickness Reynolds number ranges in the legend and will be referred as $Re_\theta$, hereafter). In this case very high Reynolds number trend is observed. Variance on the time average data for incompressible ZPG boundary layer data display that measurements beyond $Re_\theta > 7140$ have an outer peak. For tTBL study, whether intended for the flow control or high Reynolds number dependency investigation, classical threshold for optimum shear Reynolds number > 2000 was suggested by Smits et al. (2011). A slightly different value of sufficiently "High Reynolds number" estimation to observe a decade of separation between near wall and log-region coherent structures was proposed by Hutchins and Marusic (2007a) as $Re_\tau > 1700$ e.g at a critical value of $Re_\tau = 167$ inner and outer regions will have a sufficient separation.

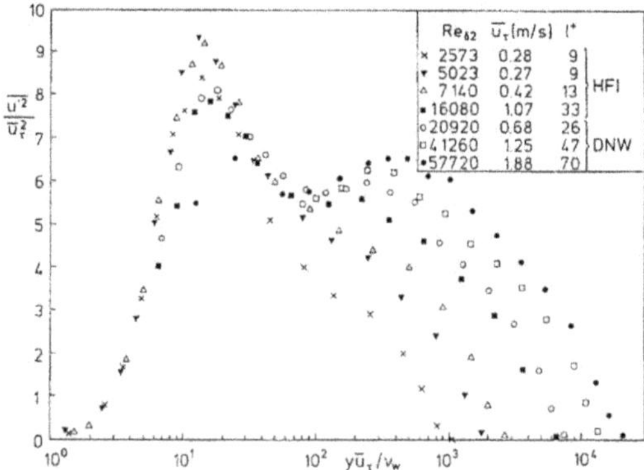


**Figure 1.11:** Streamwise turbulence fluctuation at ZPGBL (Fernholz and Finley (1996)).

Turbulent kinetic energy (TKE) production presented in semi-logarithmic wall normal distance for various Reynolds number ranges (Legend exhibits the Re ranges) from

Figure-1.12 was adapted from Marusic et al. (2010a). Authors showed that the pre-multiplied presentation of the same data indicates that TKE contribution from the logarithmic region is higher from increasing Reynolds number.

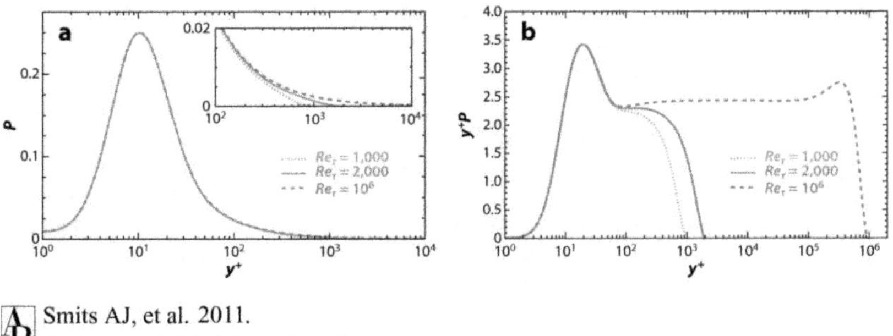


**Figure 1.12:** Turbulence kinetic energy production for a range of Reynolds numbers: (a) semi-logarithmic representation and (b) pre-multiplied representation (Smits et al. (2011)).

## 1.6 Flow Control Technique

In recent years, much emphasize has been given in the research of different flow control techniques for fluid driven high speed transportation e.g. air planes, ocean vessels, modern high speed trains and automobiles. Primary focus is towards the drag reduction as a consequence of surface friction. In United States alone, 40% drag is coming from skin friction in transportation sector whereas for a subsonic long range passenger liner, approximately 50% of the total drag is contributed from friction (Wood (2004)). Driven by the effort to reduce $CO_2$ and other Green House Gas (GHG) emission and rising fuel price, different flow control optimization techniques has been developed since the middle of $20^{th}$ century. Since 1950's, air transportation volume has exponentially increased, where subsonic passenger liners are the major fuel consumer (Banister et al. (2011)). Recent data indicate that shipping and airlines industry has spent $128 billion ($60/barrel (IMO (2015))) and $130 billion ($54.2/barrel (IATA (2017))) respectively as fuel cost. During 2018, Airline industry was estimated to spent $156 billion for fuel cost. Considering the airline cost involved in fuel expenditure, smallest saving of fuel cost determines the success/failure of the drag reduction method.

It is implausible to shift from the trend to avoid fossil fuel dependence, atleast for a foreseeable future. In order to keep the net emission at the same level from the year of 2020. International Civil Aviation Organization (ICAO) has declared a steadfast objective to improve fuel efficiency by a constant rate of 2% until 2050. In this context, an ardent challenge has been set by Airbus to reduce fuel consumption in the order of

50% within the year of 2020. Therefore, these objectives are only possible by reducing 30 - 50% of the friction drag (Kornilov (2015)). Relative financial saving by reducing drag was studied by Gad-el-Hak (1996), where a mere 10% total drag reduction of an aircraft can saves upto $ 1 billion of the annual fuel cost for the commercial airliners in USA. Under such circumstances, finding effective means to reduce friction drag is evident.

A Number of techniques has been attempted and optimized during the last half of the century and yet continues. A good review of different Flow Control Technique (FCT) can be found in the following literatures: Bushnell (1983), Gad-el-Hak and Bushnell (1991), Gad-el-Hak (2000), Lynch et al. (1991), Joslin (1998), Gad-el-Hak (2012), Choi et al. (2001), García-Mayoral and Jimenéz (2011) and NATO report (1985). Turbulent flows are inevitable, therefore, efficient modifications/control in turbulent wall bounded flows or interactions to the turbulent structures are necessary to obtain a drag reduction/flow control technique (Hough (1980)).

| management methods | Passive | Active |
|---|---|---|
| Advantage | (1) Steady state input, easy to implement.<br>(2) No external enrgy source required<br>(3) No mechanical or electro-mechanical components required | (1) Dynamic time varying.<br>(2) Adaptability to various flow conditions.<br>(3) Stand-by separation managemnet for off design performance<br>(4) Closed-loop feedback possible. |
| Disadvantage | (1) Reynolds number dependent<br>(2) Off-design sensitivity<br>(3) Often associated drag penalty | (1) Technically more complex<br>(2) External energy source necessary |

**Table 1.1:** Advantages and disadvantages of active and passive flow control techniques (Kornilov (2015)).

Based on the external energy application, control techniques are classified into two major groups as active and passive control technique (Gad-el-Hak (2000)). Different passive methods such as Riblets ( García-Mayoral and Jimenéz (2011), Vukoslavčević et al. (1991)), Microelectromechanical systems (MEMS) (Mehregany et al. (1996) and Kasagi et al. (2009)), Super-Hydrophobic Surfaces (Min and Kim (2004), Fukagata and Kasagi (2006), Daniello et al. (2009), Martell et al. (2009), Rothstein (2010) and Watanabe et al. (2017)), Polymers and Surfactants (Hoyt (1990)) and Large Eddy Break up devices (LEBU) Spalart et al. (2006) have shown optimistic performance for wall bounded shear flows but their capacity was limited for a significant drag reduction achievement. Moreover, engineering limitations imposed due to construction and maintenance perspective has restricted performances of such methods. Table-1.1 shows the advantages and disadvantages based on the characteristics of different types of flow control techniques. In the subsequent part of this section, some of the prominent flow control techniques will discussed.

### 1.6.1 Super-Hydrophobic-Surface

Super-Hydrophobic-Surface which is a special class of surface that produces large slip due to its wide contact angle to liquids. These surfaces enhance the mobility of water droplets by reducing their contact angle hysteresis. They are of similar geometry as riblets but in nano/micro scale and aligned in a transverse direction to the incoming flow. DNS performed by Min and Kim (2004) for $Re_\tau = 180$ in channel flow concluded that rate of drag reduction is increased with the increasing slip length. A theoretical description of enhanced quasi-stream-wise vortex due to SHS is reported by Fukagata and Kasagi (2006), where DNS results from $Re_\tau = 180$ and 400 was used to predict drag reduction by SHS. The asymptotic relationship of drag reduction rate to the increasing slip length of the bounding surface reaches a certain threshold for both span and stream wise slip boundary conditions. Theoretical derivation of maximum drag reduction rate of 25% was presented, although the authors did not relate such relationship based on the experimental data. As mentioned in the literature review that follows, significant drag reduction by SHS is possible only within air-water boundary and found effective for hydrodynamic applications.

In a more elaborated study, continuous air injection in a gas-liquid interaction was extensively studied experimentally, where authors have measured in excess of 80% drag reduction of such technique of TBL (Elbing et al. (2008)). Experiments using SHS is quite effective for hydrodynamic applications and can obtain significant drag reduction. Contrary to the technique where air-air active control is applied for aerodynamic applications, SHS is more focused to the hydrodynamic research for gas-water interaction for passive control (Rothstein (2010)).

### 1.6.2 laminar flow control

On the contrary, active methods which are a function of external energy supply has shown rather superior performance to reduce friction drag. Some of those are active laminar flow control (LFC) (Joslin (1998)). One of the outstanding technique is LFC which was started developing since 1930s and was being developed for next 70 years subsequently.

With this technique, suction through a porous surface is used to delay the boundary layer transition in order to reduce the cumulative drag over the entire surface. This was found effective reducing skin friction drag by keeping a large part of boundary layer in laminar regime. Previously it was believed that such a method with its varying location of suction location can significantly reduce drag (Joslin (1998)) but recent DNS result from Kametani and Fukagata (2011) has shown that suction increases local drag but reduces local turbulence intensity. A narrative on the applied aspect of such control mechanism can be found in Joslin (1998) and Stroh et al. (2016).

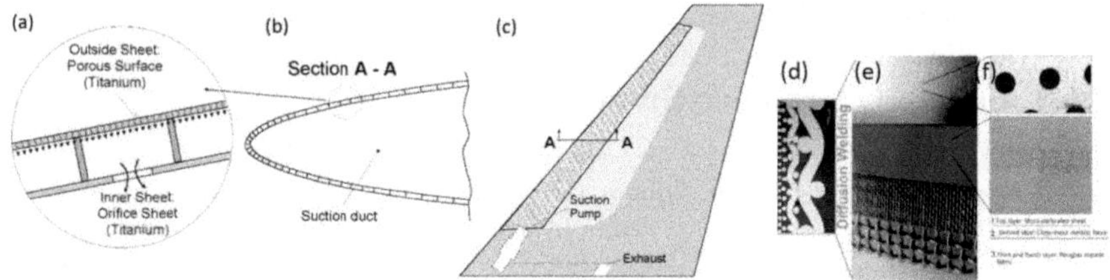


**Figure 1.13:** (a), (b) and (c) ALTTA Micro-perforated skin schematic for HLFC surface implemented on the leading edge of Airbus A320 tail section developed by German Aerospace Center (DLR) (ALTTA stands for Application of hybrid Laminar Technology to Transport Aircraft) Beck et al. (2018); (d) Segment of tail section of Boeing A320 with simplified micro-perforated surface with single suction chamber; (e) Different layer of surfaces beneath the micro-perforated surface for HLFC; (f) Krishnan et al. (2017).

LFC falls within the category of active flow control technique utilized in aviation industry in order to maintain laminar state of the flow at chord Reynolds numbers beyond which the flow will normally be considered as being transitional/turbulent in the absence of control. Relaminarization of a turbulent flow state is not the same as the laminar flow control. Therefore, it is often misinterpreted as a 'relaminarization' process although both flow physics phenomena may apply the same control system. Depending on the type of surfaces of a flying body (fuselage, wings etc.) such control mechanism may be applied principally in two different ways. In a way the 'Natural Laminar Flow' (NLF) is applied by creating an artificial favorable pressure gradient over the surface to delay the natural transition process to turbulent zone. Most often NLF fails to achieve a required performance for drag reduction due to the formation of inherent boundary layer instabilities. In order to overcome the limitations of LFC, a 'Hybrid Laminar Flow Concept' (HLFC) was introduced in order to reduce the suction requirements in a wide area of control surface and therefore, can reduce the system complexity by applying suction only in the narrow region of the leading edge of the wing. Despite significant challenges, HLFC technology is in the most matured state of development.

Very recently, real scale flight test from Airbus A320 transport aircraft in 1998 was found aerodynamically successful, where micro-perforated surface was used to implement HLFC through uniform continuous suction. However, from structural point of view, suction surface required further development for simplification. Additionally, long term use of perforated surface for suction may have insect contamination and will cause financial penalties from the maintenance perspective (Corda (2011)). In order to get the detailed view on the topic, Joslin (1998), Brasslow (1999), Bushnell (2003) and Reneaux (2004) are advised for interested readers.

In order to overcome these imperfections, subsequent flight test in 2017 using an Airbus A340-300 (Airbus Press Release (2020)), which included HLFC through perforated surface not only limited to the tail leading edge but also to the wing leading edge. These are the current development for active flow control techniques using perforated surface and exhibit the readiness of the Technique.

### 1.6.3 Large Eddy Break-up Devices

One of the burning issue regarding the evaluation of a particular flow control technique in SBL is the extent of the effected region both in vertical and longitudinal direction. With increasing Reynolds number, a persistent effect is desired when control is applied. Therefore, one of the simplest way to effect the large structures beyond the viscous sub-layer is a parallel plate placed upstream. This is done in order to break the large eddy structures and to achieve a desired drag reduction effect simultaneously.

Parallel plate manipulator for the larger eddies are also known as 'Large Eddy Break-up Device' (LEBU) or parallel plate manipulator. Preliminary results from Corke et al. (1981) using Hot Wire Anemometry (HWA) measurements exhibit the damping of streamwise velocity fluctuation. Measurements taken at considerably low Reynolds number displayed persistent effect upto $70\delta$. Their experiment was found very effective to inhibit the intermittent large-scale structures of the SBL. Although they did not confirmed a net reduction of the skin friction. Later, Direct Numerical Simulation results from Spalart et al. (2006) rejects the idea of LEBU using for aerodynamic surfaces. Although, their finding was very interesting as LEBU devices can effectively break-up the larger eddies into smaller ones but streamwise turbulence production is very quickly recovered as opposed by the findings from Corke et al. (1981). Recently, Chin et al. (2017) performed Large Eddy Simulation (LES) of a spatially developed zero-pressure-gradient TBL within the range of $Re_\theta = 500 \sim 4300$. Persisting effect of LEBU devices were found to be active upto $160\delta$ downstream, velocity deficit of the wake region downstream of a LEBU diminishes gradually with streamwise distance. Similar to the study from Corke et al. (1981), results from Anders (1989), Spalart et al. (2006) and Chin et al. (2017), no net reduction of friction drag was confirmed. Therefore, further literature survey on the method have shown very little/no net reduction of the skin friction drag so far.

### 1.6.4 Riblets

In a different approach that was first introduced by Liu et al. (1966), later followed by several other investigation such as Vukoslavčević et al. (1991), García-Mayoral and Jimenéz (2011) and very recently by Spallart and McLean (2011)). Riblets are similar to rough walls, is a surface of grooves aligned to the mean flow direction. Though it requires no external energy but increases the wetted surface to planform area ratio and subjected to re-installation every 5 years. Thus, such FCT is not economically feasible with a maximum of 15% of friction drag reduction.

### 1.6.5 Other techniques

Several other control techniques that are extensively studied to control SBL are jet actuators (Choi et al. (2001), Choi (2001) and Mahfoze and Leizet (2017)), Opposition control (Kim et al. (2003), Stroh et al. (2015) and Abbassi et al. (2017)), Microelectromechanical systems (MEMS) (Kasagi et al. (2009)), Polymer additives (White and Mungal (2008) and Benzi (2010)) and gas microbubbles ( Legner (1984) and Merkle and Deutsch (1989)) were found with positive outcomes. More recently blended wing body (BWB) (Ko et al. (2003)) and boundary layer ingestion (BLI) (Smith ET AL: (1993) and Plas et al. (2007)) are attractive concepts presently under investigation in aviation industry.

Among all control methods, blowing with air or other gases with different viscosities has the potential to alleviate surface friction in access of 50 % (Hwang (2004)). Therefore, within the context of this paper, we will focus on the experimental investigation of the uniform blowing as a mean to Turbulent Drag Reduction (TDR) and thereafter, its consequence on the turbulent boundary layer.

Reduction of drag is one of the principle factors that is directly influencing aircraft efficiency which is also in turn, enhance the range, speed and payload, reduce operating cost and GHG emission. Other factors such as aircraft engine and shape has been significantly optimized for last decades but much can be done in order to reduce the drag. In fact, skin friction reduction within incompressible shear flows is considered a major "Barrier problem" to the further optimization of the most aerodynamic and hydrodynamics bodies (Bushnell (1983)). Classical aerodynamics conveniently segregated the total drag into pressure or form drag that include interference and roughness drag, lift drag, compressibility drag and drag due to viscosity which is also known as skin friction drag. Thereby, based on the extent of boundary layer over a subsonic aircraft, now we know that laminar region is considerably smaller than the turbulent region. Therefore, exploring viscous drag reduction in and around turbulent zone is one of the major opportunity where substantial reduction will effect the net drag contribution. Eventually, this will lead towards the overall fuel savings.

### 1.6.6 Micro-blowing Technique

Air blowing also known as injection or constant mass flux or Micro-blowing Technique (MBT). In the previous section, TBL researches were discussed. What follows in this section, concerns different research initiatives which are conducted to explore an effective way to apply air blowing. Air blowing through slots of the subsonic wing was investigated by (Schlichting (1942a) and Schlichting (1942b)), where additional energy is added in upstream control region to relegate separation at high angle of incidence for favourable pressure region. Later, a succession of experimental studies refined the empirical aspects of blowing technique. Large array of incompressible TBL data in ZPG condition was generated using wind tunnel experiments. Majority of the researchers developed empirical correlations from their experimental data taking Prandlt's (Prandlt (1927)) mixing length theory as basis where porous boundary condition and wall normal blowing effect was implemented. Jeromin (1970) summarized the sequence of these experiments and suggested that comprehensive measurements at high Reynolds numbers are required in order to derive the physics of the flow. In addition to experiments, analytical derivation of such flow is found from Catheral et al. (1965) but the authors were unable to show a satisfactory agreement with corresponding numerical solutions.

One major issue regarding porous surface is the growing skin friction compared to the smooth surfaces in TBL. This has been investigated by Townes and Sabersky (1966), Burden et al. (1970) and Wilkinson et al. (1988) at low Reynolds number TBL. As a result of injected air, the used porous surface is subjected to increased drag phenomena where porosity, aspect ratio and shape of the holes are the determining factors. Burden identified the increased drag phenomena without blowing, indicating two factors being responsible, first destabilization due to the rapid creation of adverse and favourable pressure region adjacent to the wall, second destabilization from the vortex flow pattern within the holes itself. Wilkinson proposed that the blowing can augment this process up to 60 wall units for the flow scale of $Re_{\theta,SBL}$=1000 and speculated more violent bursting events that will persist further away from the wall as flow Reynolds number increases. However, such assumption is yet to be proven.

In order to apply wall normal blowing, selection of the blowing surface is an important decision. Therefore, several factors have to be taken into account to select a near ideal perforated surface. According to Gregory (1961), although the ideal porous surface does not exist since the normal velocity at the boundary must be zero on the surface between the holes/pores.

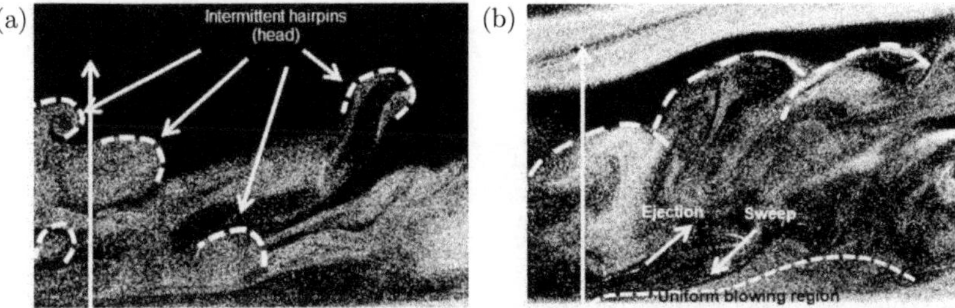


**Figure 1.14:** (a) Flow visualization over smooth surface TBL in the field of view (FoV) as indicated in Figure 11 at $Re_\theta \approx 1100$ using smoke ; (b) Flow visualization over perforated surface with uniform blowing at 0.7%; the vertical arrow indicate the boundary layer thickness and flow is coming from left to rights from readers reference for both (a) and (b);

Although not necessarily the most practical, selecting a material whose holes and holes spacing are both small compared with the boundary layer thickness is the nearest approximation to the ideal surface selection. Despite highly 3D inflow pattern, even blowing through discrete perforations in a solid surface has been found to be effective in order to maintain boundary layer attached to the wall under certain conditions. The permissible distribution of such perforations, whether uniformly over the surface or concentrated into narrow spanwise slots, has therefore, also been a subject for research. In discussing these matters in the sections that follow it is useful to recall the approach to the ideal surface, summarized in Figure-1.15. This divides the possibilities into two groups according to whether the inflow is 2D or 3D. TBL is of highly 3D in nature, therefore, according to Figure-1.15, a surface with uniformly distributed discrete perforations/holes is the best choice to apply for flow control in TBL.

There exist two principle aspects for effective applications of MBT, first, an optimum surface and blowing rate as a function of external energy, second, modification to the flow. First reported study on the perforated surfaces can be found from McQuaid (1968). Very recently, first aspect has been extensively investigated in a series of experiments from National Aeronautics and Space Administration (NASA), Glenn Research Center, notably Hwang (1997) compared several MBT skins with a wide variation of injected air. Figure-1.16 (a), (b) and (c) presents some of the earlier perforated surfaces from McQuaid (1968). In recent days, Figure-1.16 (d) and (e) are more common types of perforated surfaces used for MBT application from Hwang (2004) and Hasanuzzaman et al. (2021) respectively.

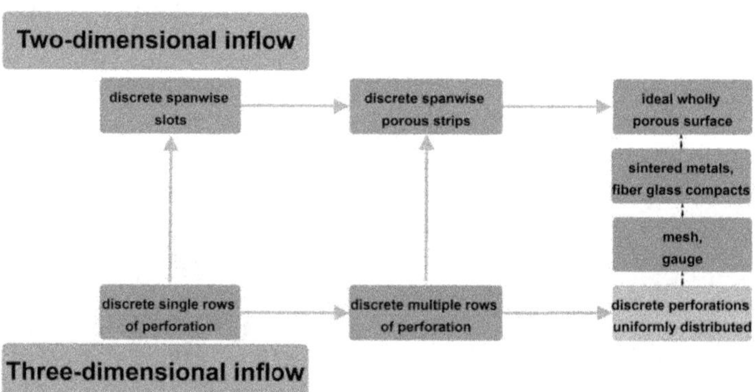


**Figure 1.15:** The approach to the selection of near ideal perforated surface. Reproduced from Gregory (1961)

Force balance measurements of skin friction at compressible ZPG-TBL were used to develop an optimum MBT surface. The experiment was successful to optimize the geometry of a MBT surface considering inclination angle, pattern, arrangement, diameter, porosity and aspect ratio. Though the experiments optimized the technique but were unable to provide any explanation to the change in turbulence parameters. MBT surface optimization has been reviewed in details from Hwang (2004).

The advantage of MBT is that the finite length of affected area through MBT can be persistent for the complete spatial growth of TBL (Stroh et al. (2016)). Thus, according to Kornilov (2015), very small amount of external energy input in the form of MBT can effectively alter several boundary layer and turbulence properties with relatively simple but robust construction. Moreover, as an outcome, in excess of 50% reduction is possible for skin friction drag (Hwang (2004)). As such, chronological development of MBT in details will be discussed in the subsequent part of this section.

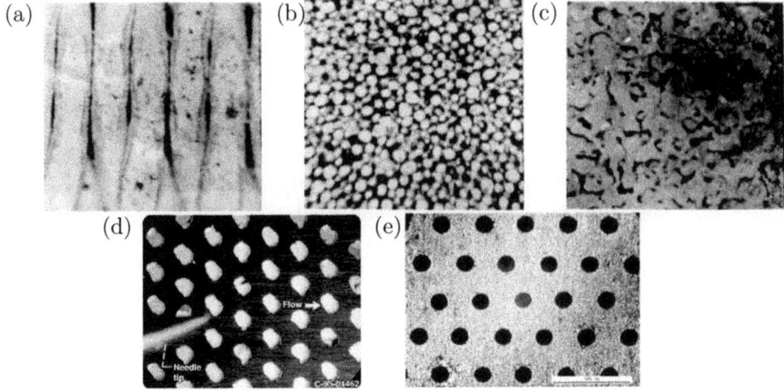


**Figure 1.16:** (a) Poroloy stainless steel wiremesh, (b) Porosint, (c) Vyon, (d) Laser drilled stainless steel sheet and (e) Electron beam drilled stainless steel.

### 1.6.7 Legacy of TBL with transpiration/perforation

Apart from the early paper by Schlichting (1942a) and Schlichting (1942b), at different times the technique of blowing in conjunction with suction was termed in different names such as transpiration, injection and blowing. To avoid perplex understanding of the control technique, in the following text we will use blowing instead. The flow control method discussed in the scope of this proposal is confined within the framework of TBL over flat plate in zero pressure gradient condition where the bounding surface at certain stream-wise distance will be replaced with permeable/perforated surface instead of smooth surface. Schlichting (1942a) and Schlichting (1942b) showed the relationship between the asymptotic boundary layer under the influence of uniform blowing comparing the momentum thickness in laminar and TBL. Later, all the investigations were based on linear mixing length where the effort rested on to relate turbulent shear stress to the local velocity gradient ($\partial u/\partial y$) in the laminar sub-layer.

#### Mickley and Davis

Dawning of the experimental investigation for boundary layer flow control with blowing and suction started as early as forties in the last century. Contrary to the other flow control method, air injection/uniform blowing carried an added advantage of cooling to the boundary layer flow with thermal gradient. Least, compared to boundary layer separation control with suction, blowing was used for drag reduction and cooling effects. A preliminary model of TBL with blowing and suction was presented by Mickley and colleagues Mickley et al. (1954), where the authors concluded that blowing increases the boundary layer thickness and reduce the magnitude of surface friction. However, a qualitative model based on the film theory was presented as the blowing air acts as a layer analogous to a developing film in the downstream locations but did not succeeded for the prediction of friction co-efficient. In a similar study of transpiration (Blowing and suction) in compressible TBL reported by Rubesin (1954) concluded that the blowing greatly decreases friction drag as well as increases heat dissipation from the surface. A subsequent investigation by Mickley and Davis (1957) came up with a more complete and progressive extension to their previous work and derived a logarithmic relations for the profile measurement taking wall normal blowing velocity in consideration. The logarithmic relationship for laminar sub-layer and turbulent region (log-layer and wake region) were separately treated and expressed in the following form.

Equation-1.36 shows their prediction of the velocity profile where wall distance is predicted as a function of measured wall shear velocity and blowing velocity.

Equation-1.37 presents the expression for inner layer as, $0 \leq y^+ \leq y^+_{inner}$. Where, $y^+_{inner}$ denotes the end of the laminar sub-layer and superscript $^+$ denotes the non dimensional wall normal distance.

Mixing length Constant, $\kappa$ was determined using different blowing velocity from wall and was increasing with increasing blowing ratio, $V_w/U_\infty$. The local skin friction co-efficient was obtained using this relationship from Equation-1.37, where blowing velocity ($V_w$), local shear velocity ($u_\tau$), mixing length constant ($\kappa$), local non-dimensional wall normal height of laminar sub-layer ($y^+_a$) and local stream-wise velocity component at, $u = u_a$.

$$y^+ = \frac{2u_\tau}{V_w} ln\sqrt{(1 + \frac{V_w}{u_\tau})} \tag{1.36}$$

$$ln\frac{y^+}{y^+_a} = \frac{2\kappa u_\tau}{V_w}\left(\sqrt{\left(1 + \frac{V_w \bar{u}/u_\tau}{u_\tau}\right)} - \sqrt{\left(1 + \frac{V_w \frac{U}{u_\tau}\Big|_{y=a}}{u_\tau}\right)}\right) \tag{1.37}$$

However, friction co-efficient ($C_f$) could not be related satisfactorily to the Reynolds number based on plate length ($\mathrm{Re}_x$). On the other hand, local Reynolds number based on momentum thickness ($Re_\theta$) and skin friction co-efficient relationship was expressed by the following Equation-1.38.

$$ln\frac{\kappa Re_\theta}{y^+_a\left(1 + \frac{V_w}{u_\tau\sqrt{C_f/2}}\right)} = \frac{2\kappa u_\tau}{V_w}\left[\sqrt{\left(1 + \frac{V_w}{u_\tau\sqrt{C_f/2}}\right)} - \sqrt{\left(1 + \frac{V_w u^+_a}{u_\tau}\right)}\right] \tag{1.38}$$

They performed their experiment over a wide range of moderate Reynolds number based on momentum thickness ($1000 \leq Re_\theta \leq 7000$) and uniform blowing $0 \leq V_w \leq 0.5$ % the conclusion drawn at Mickley et al. (1954) was further corrected in the preceding Mickley and Davis (1957) and 15 30 % skin friction drag reduction was claimed with uniform blowing depending on various blowing rate. However, buffer layer which is the intermediate transition between the laminar sub-layer and logarithmic zone was not taken into consideration. Moreover, uncertainties regarding wall friction measurement at higher blowing rate were recommended to measure to observe a stochastic boundary layer under the influence of uniform injection/blowing.

#### Rheinboldt

Rheinboldt (1956) studied the suction/blowing boundary layer when suction/blowing was applied over a finite length of the plate. He discovered that the boundary layer separates beyond a specific downstream length $x = 0.7456 U_\infty \nu / V_w^2$ , when blowing is applied. For strong blowing, the boundary layer separates earlier than for small blowing velocities.

#### Turcotte and Leadon

After taken into account the buffer layer treatment, Turcotte (1960) developed a theoretical velocity profile which had a good agreement with the data from Mickley and Davis (1957). He concluded that the extent of mixing length ($l_m$) is limited only up to the region of buffer layer at low blowing rate and suggested a similarity parameter ($V_w/u_{\tau,SBL}$) , here $u_{\tau,SBL}$ correspond to the wall shear velocity obtained in the reference case of smooth wall) to relate blowing velocity. However, the theoretical velocity profile failed to quantify shear stress accurately. In response et al. (1961) proposed a new similarity parameter in connection with free stream velocity. Details can be found in the review of Craven (1960). A conglomeration over a wide range of skin friction data followed by summary discussion can be found in the following paper. Craven compiled data of skin friction coefficient over boundary layer flow in different boundary conditions under the influence of blowing and suction and concluded that blowing from permeable surface reduces skin friction coefficient in all kinds of boundary layer flow compared to the flow over impermeable surface (Incompressible laminar, incompressible turbulent, compressible laminar and compressible turbulent).

#### Catherall, Taylor, Richardson and Hokenson

Thus in a separate attempt to define the boundary conditions for flow with uniform injection, Catheral et al. (1965) provided a mathematical model. However, they were successful to define boundary conditions over permeable surface but could not provide a satisfactory agreement between the numerical model of the equation of motion and continuity to their analytical solution. A nice review about modeling of boundary conditions of the flow over impermeable/perforated surface can be found from Taylor (1971), Richardson (1971) and Hokenson et al. (1985).

### Stevenson

Widely used law of wall for zero pressure gradient TBL with uniform blowing from permeable/porous wall was proposed by Stevenson et al. (1963). Contrary to Mickley and Davis (1957), Pressure gradient term in the mean flow was validated in the mean momentum equation and a subsequent law of wall relationship was proposed with a common values of Mixing length constant ($\kappa$) and integration constant B (Equation-1.41).

$$\frac{2u_\tau}{V_w}\left\{\sqrt{\left(1+\frac{V_w\bar{u}}{u_\tau^2}\right)}-1\right\}=\frac{1}{\kappa}ln\frac{yu_\tau}{\nu}+B \tag{1.39}$$

$$\frac{1}{V_w/u_\tau}\left\{\sqrt{\left(1+\frac{V_w}{u_\tau}\frac{\bar{u}}{u_\tau}\right)}-1\right\}=\frac{1}{\kappa}ln\frac{yu_\tau}{\nu}+B \tag{1.40}$$

$$\frac{2}{V_w^+}\left\{\sqrt{(1+u^+V_w^+)}-1\right\}=\frac{1}{\kappa}lny^++B \tag{1.41}$$

Theoretically, a slight increase in blowing velocity can cause a large amount of skin friction coefficient reduction ($C_f = 2u_\tau^2/U_\infty$). Stevenson et al. (1963) determined the value of $\kappa$ and A from his experimental data e.g. 0.4 and 5.8. Nevertheless, values obtained in the present thesis for constant of integration ($A = 5.3$) differs from that of Stevenson et al. (1963) but mixing length constant ($\kappa = 0.4$) was found in agreement. This difference or scatter of the integration constant is apparently due 10 low Reynolds number effects.

### Black and Sernecki, McQuaid and Tennekes

Based on the Equation-1.1, Black and Sarnecki (1965), extended the logarithmic law for the impermeable, smooth surface to constant wall normal blowing from permeable surface. Thus they proposed a bi-logarithmic law obtained from the Momentum transfer theory and extended the Coles (1956) with the same boundary condition. Their mathematical expression for the Bi-logarithmic law and wake law for impermeable surface with blowing is as follows:

$$\frac{\overline{u}}{U_\infty}-\frac{1}{4\kappa^2}\frac{V_w}{U_\infty}\left(\ln\frac{U_\infty y}{\upsilon}\right)^2=\left[\frac{1}{4\kappa^2}\frac{V_w}{U_\infty}\left(\ln\frac{U_\infty\delta}{\upsilon}\right)^2-\frac{u_\tau^2}{V_wU_\infty}\right]-\left(\frac{1}{2\kappa^2}\frac{V_w}{U_\infty}\ln\frac{U_\infty\delta}{\upsilon}\right)\ln\frac{U_\infty y}{\upsilon} \tag{1.42}$$

$$where,\qquad \frac{V_w}{u_\tau}=\frac{1}{\kappa}\Pi\left(x\right)\omega\left(\frac{y}{\delta}\right) \tag{1.43}$$

Here, second part of the Equation-1.43 is derived from the Coles wake law (Coles (1956)) and $\Pi$ is the profile parameter, $\omega$ is Cole's Wake function and $\delta$ is local boundary layer thickness.

From their experimental value they have concluded that the assumptions of momentum transfer theory with linear mixing length hold true even in the case of blowing and included the wall normal blowing velocity in the mixing length equation, as such the Equation-1.44 was derived.

$$\left(\kappa y \frac{\partial u}{\partial y}\right)^2 = u_\tau^2 + V_w \bar{u} = \frac{d}{dx}\left(U_\infty{}^2\theta\right) - V_w(U_\infty - \bar{u}) \tag{1.44}$$

McQuaid (1968) performed experiment under incompressible TBL varying the blowing rate. A detailed description of the porous bounding surface was discussed and various ways to distributed porous surface to influence mean momentum in the boundary layer was presented. Homogeneity in terms of uniform lowing was discussed in the literature of McQuaid (1968). Tennekes (1965) criticized the bi-logarithmic formula developed by Black and Sarnecki (1965) due to its weakness being universal in the laminar sub-layer. According to his argument, further experiments with higher blowing ratio are required to modify Equation-1.43.

## Simpson, Rotta and Jeromin

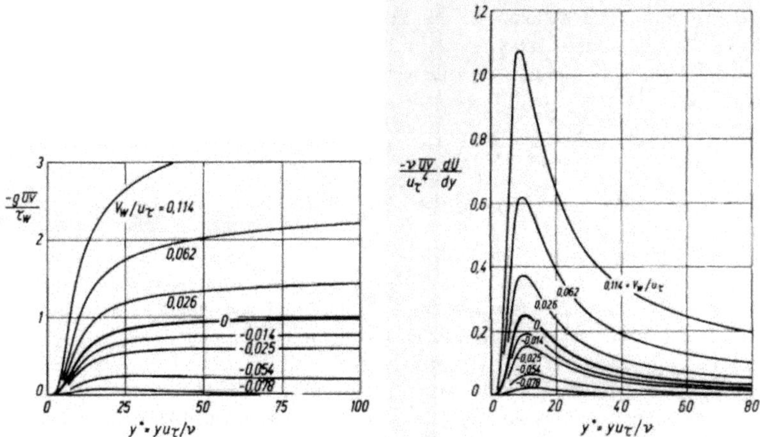


**Figure 1.17:** (a) Inner scaled Reynolds Shear Stress and (b) Turbulent Energy Production along wall normal location for different blowing rate expressed as $V_w/u_\tau$.(Rotta (1970))

Simpson and colleagues presented experimental data primarily focused on the skin friction data over a variety momentum thickness Reynolds number range. Conclusion drawn from their validation to the former studies has found more similarity to the theory developed by Rubesin (1954) in terms of friction measurement with varying blowing.

A comparison between the Coles's wake law hypothesis by Black and Sarnecki (1965) was re-evaluatted by Rotta (1970). An elaborate presentation of the analytical formula for Reynolds shear stress at varying blowing was incorporated with the blowing velocity ($V_w$). In addition he was also able to produce accurate measurements for the turbulent energy production in the laminar sub-layer where viscosity is dominant. To calculate Reynolds shear stress and turbulent kinetic energy analytically, Rotta used Equation-1.46 which is presented in Figure-(1). Jeromin (1970) reported a comprehensive review about compressible and incompressible TBL with blowing effect.

$$-\overline{uv} = (\kappa y)^2\left[1 - exp\left\{-y\sqrt{u_\tau^2 + V_w\bar{u}/\nu A}\right\}\right]\left|\frac{d\bar{u}}{dy}\right|\left|\frac{d\bar{u}}{dy}\right| \tag{1.45}$$

$$\frac{d\bar{u}}{dy} = \frac{2u_\tau^2 + V_w\bar{u}}{\nu + \sqrt{\nu^2 + 4\kappa^2y^2\left[1 - exp\left\{-y\sqrt{u_\tau^2 + V_w\bar{u}/\nu A}\right\}\right]^2(u_\tau^2 + V_w\bar{u})}} \tag{1.46}$$

followed by Rotta (1970). In the later scientific publication the author compared Bradshaw's shear stress transport equation and A. M. O. Smiths eddy viscosity relation with the experimental data obtained in equilibrium boundary layer under the effect of suction and air injection. Theory was developed in conjunction with the Coles law of wake (Coles (1956)) and boundary conditions with transpiration was obtained related experiment was conducted over a wide range of stations in a TBL using surface pitot tube. Experimental data was used to develop an empirical formulation of the turbulent boundary layer equations with transpiration.

#### Computational Fluid Dynamics

Numerical simulations have been extensively used to investigate MBT in wall bounded shear flows. LES results from Piomelli et al. (1989) at $Re_{\theta,SBL}$=1195, showed that blowing thickens the turbulent channel flow on the control side (over the blowing surface), reduces skin friction and increases turbulence intensity. Moreover, turbulence spectra indicate that blowing increases the energy content of the small scales which is also identical from the recent results discussed in the later section.

Sumitani and Kasagi (1995), performed DNS on TCF geometry at $Re_\tau, SBL = 150$ where they have applied blowing ratio (BR)[1] at 0.00344 %. Their imperative regarding the low pressure regions was well responsive to the coherent turbulent vortex cores which generates high Reynolds stress events (Ejections and sweeps). Therefore, they concluded that the blowing stimulates manifestation of the coherent streamwise structures resulting in increased production and Reynolds stresses.

Particularly, study by Kametani and Fukagata (2011) is of great relevance in terms of data validation and to the further outline of this thesis. Second invariant of deformation tensor from DNS data at $Re_\theta = 530$ was used. Finally, the spatial development of the TBL reaches to a maximum of $Re_{\theta,SBL}$=700. Flow visualization and identification

[1] Here, $BR = U_b/U_{bulk} \times 100$ and $U_{bulk}$ is the bulk mean velocity of the channel.

of vertical structure in outer region was done using iso-surface criteria and displayed in Figure-1.18. In addition, color scheme on the wall indicate the wall shear. Figure-1.18(top), indicates the TBL over smooth surface, on the other hand bottom figure indicate flow over perforated surface at uniform blowing rate for 1% of $U_\infty$. The flow visualization in outer region using iso-surface criteria, showed an increased population of large eddies when subjected to uniform blowing.

$$\frac{-\bar{u}V_w}{-\overline{u'v'}} = \frac{V_w}{U_\infty} ln^2(Re_\tau) \tag{1.47}$$

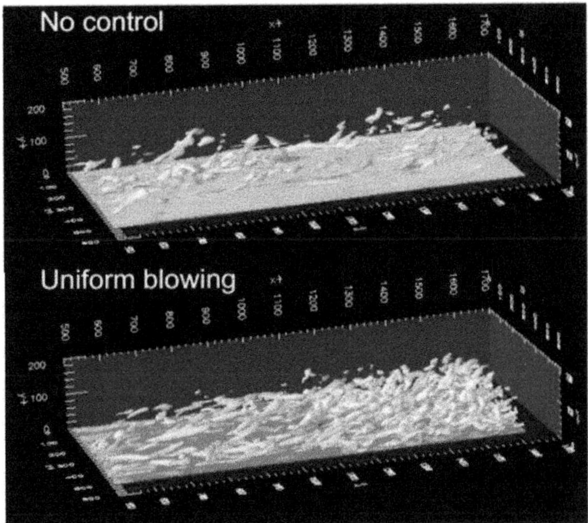


**Figure 1.18:** Flow visualization of coherent structure using second invariant of the deformation tensor ($Q^{(+0)}$ or iso-surface criteria), colour scheme on the surface indicate local wall shear ($\tau^{(+0)}$; (top) ), (bottom) uniform blowing at BR = 1% (Kametani and Fukagata (2011)).

The effect of blowing was identified as a reasoning for drag reduction and vorticity enhancement simultaneously. Most importantly, they estimated using Equation-1.41, which is the derived equation of log law for wall with blowing by Stevenson et al. (1963), here usual notation is used except $V_w$ which indicate dimensional blowing velocity (positive in wall normal direction). They concluded with an assumption that, the drag reduction effect may increase for the cases at the higher Reynolds number flow provided that the blowing magnitude is in the same order e.g Equation-1.47 states that if the blowing amplitude is constant with respect to the free stream velocity, i.e. if $V_w/U_\infty$ is constant, the drag reduction effect is expected to be stronger at higher Reynolds number.

However, Reynolds number of DNS results from Kametani and Fukagata (2011) was way to smaller in comparison to the realistic engineering application and do not include a substantial picture of the drag reduction mechanism.

Kametani et al. (2015) investigated the effect of uniform blowing on large scale structures for TBL flow at $Re_{\theta,SBL}$ = 2500 using Large Eddy Simulation (LES). Where, BR was varied at three different rates e.g. 0.1 %, 0.5 % and 1 % of $U_{\infty}$. Premultiplied spanwise power spectra of streamwise velocity clearly shows the impact of blowing on both the inner and outer layer, while increasing the range of wavelengths in both regions. In presence of blowing, the outer region is even more prominently influenced, as the premultiplied co-spectra (streamwise and spanwise) reveals formation of a second peak at $y^+$ = 100. This phenomenon shows on one hand that the large scale structures are enhanced, and on the other hand proves the increased contribution of large scale structures to skin friction drag. Due to the limited Reynolds numbers and blowing ratios in similar studies, it is not possible at this stage to formulate a comprehensive hypothesis on how the structures react to various blowing ratios. Principle findings are summarized as follows:

1. Turbulent Kinetic Energy (TKE) is increased by blowing.
2. Blowing increases the small short wavelength component of streamwise and wall-normal component.
3. Near wall production is increased.
4. The range of the wavelength and wall distance of the spectra is spread by blowing.
5. Inner peak energy is increased by blowing but the position is mostly unaffected.
6. Indicate that inner and outer regions are differently affected.
7. Effect of the blowing is more pronounced in the outer layer.
8. Contribution from the Reynolds Shear Stress (RSS) term is affected by blowing.
9. The turbulent fluctuation is enhanced by the wall flux induced from the wall.
10. Second peak in the pre-multiplied spectra for streamwise velocity is more prominent when blowing was applied, in turn this indicates more coherence and elongation of the large scale structures in the outer region.

Recent DNS results at relatively higher Reynolds number for ZPGTBL from Stroh et al. (2016) compared two flow control schemes namely, body force damping which is quite similar to suction and uniform blowing from a controlled region in upstream location. Spatially developed TBL correspond to $Re_{\theta,SBL}$ = 2500 where blowing was applied at $Re_{\theta,SBL}$ = 470 ~ 695 (Pink area in Figure-1.19) with a magnitude of BR=0.5%. Results show that persistent spatial stretch of such a low magnitude blowing reaches far downstream to the entire length of computational domain and higher population of outer layer structures. Figure-1.19 (b) and (c) present the spatial TBL over smooth and blowing surface respectively, where, distinct growth of vorticity and turbulence is visible on the outer layer.

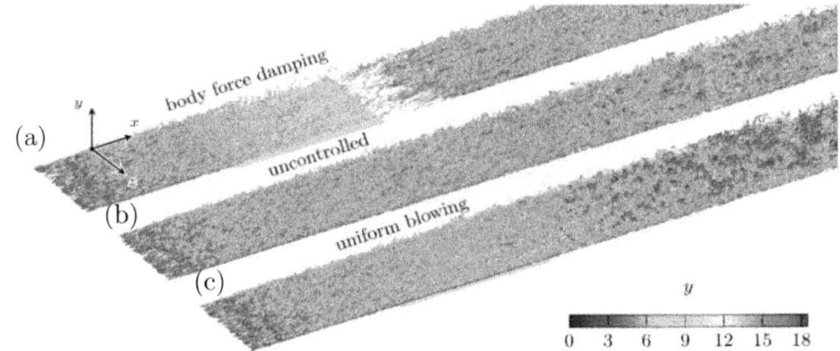


**Figure 1.19:** Iso-surfaces of $\lambda_2$ criterion coloured by wall normal coordinate. Red shaded area marks the control location (Stroh et al. (2016)).

Figure-1.20 presents the LES results of Atzori et al. (2020) at $Re_x$=0.2 million over the surface of an asymmetric NACA4412 airfoil. MBT is the FCT developed for the aerodynamic surfaces, They have used both suction and blowing on different sides of the said airfoil with different combinations. Such as the location of blowing and suction. They have recommended that suction side of the airfoil is not suitable for application of uniform blowing. It increases the pressure drag and ultimately leads to higher total drag despite local skin friction is reduced. Moreover, it decreases lift. Although, small modifications of the pressure drag and lift is observed, they have recommended to apply uniform blowing on the pressure side of the airfoil as it increases aerodynamic efficiency in addition to reduction of skin friction and total drag. Other significant findings are as follows:

- uniform blowing has effects on the turbulent statistics that are similar to those of adverse pressure gradients[2].
- blowing enhances the wall-normal convection.
- It is critical to understand to what extent the results of studies such as the present one are relevant at higher Reynolds numbers, closer to practical applications.
- uniform blowing over the suction side will always increase the pressure drag.

[2]In order to clearly distinguish the difference between the adverse pressure gradient (APG), ZPG and favourable pressure gradient (FPG) TBL over smooth surface, Harun (2011) presented their results using HWA. Their results indicate that the energy in the outer layer increases as the pressure gradient increases. The so called LSM are much more energetic for adverse pressure gradients compared with the other cases. Using uniform blowing at ZPG also energizes the outer layer energy contribution, hence, one can also observe this through turbulent statistics

- final recommendation:uniform blowing applied over the pressure side.

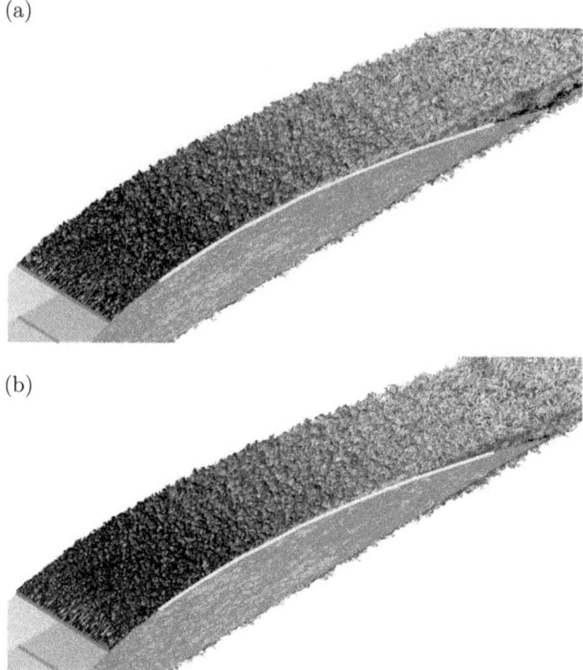


**Figure 1.20:** Outer layer vortex where, vortex clusteres are coloured with the instantaneous velocity component, from (red) $u \approx 1.7$ to (blue) $u \approx -0.2$ . The yellow and red lines indicate the spanwise controlled region and the tripping location, respectively; (a) uniform blowing and (b) uniform suction (Atzori et al. (2020)).

They have used uniform blowing and suction on airfoil location $0.20 < X/C < 0.86$ at BR≤0.2%. For both blowing and suction cases suction side was used. Figure-1.20 (a) and (b) exhibit the effects of uniform blowing and uniform suction respectively over the instantaneous streamwise component of the velocity. Where, vortex clusters were identified with the $\lambda_2$ criterion (Jeong and Hussain (1995)). The effect of blowing and suction is apparent in the vicinity of the trailing edge. However, authors remained critical about the understanding to what extent the results of this study is relevant at higher Reynolds numbers, closer to practical applications.

Numerical simulations can provide details of the flow over entire domain but they are still restricted to the computational capacity to reach certain Reynolds number that is necessary as indicated by Hutchins and Marusic (2007a). At high Reynolds number beyond the threshold value (e.g $Re_\tau \geq 1700$), numerical data is largely unavailable and necessitates further experimental study. Furthermore, the experimental and numerical studies previously discussed, provide a comprehensive understanding of how MBT modifies the flow, but they fail to present a solid theory on how the modifications are caused.

Hasanuzzaman et al. (2016) reported about the mean profiles at low Reynolds number TBL flows, drag reduction of 13% was found at $\mathrm{Re}_{\theta,SBL}$=1788 with uniform wall normal blowing. However, profiles of streamwise velocity under the influence of wall normal blowing requires different scaling other than logarithmic or power law. Moreover, high Reynolds number behaviour of the time averaged data was necessary in order to verify that the similar impact is also observed when blowing is applied. As a legacy to the the previous study at moderate Reynolds number, Hasanuzzaman et al. (2018) reported instantaneous flow field data from time resolved SPIV measurements at a spatially developed high Reynolds number TBL. They reported on experiments investigating enhanced outer layer vortices at $Re_{\theta,SBL} = 7495$, spectral results showed that blowing is expected to add energy to the streamwise velocity component as an active method. This happens as the addition of energy is dependent on the blowing ratio where wall normal component is expected to curtail the magnitude of the streamwise velocity. Eventually, changing the mean gradient $(du^+/dy^+)$ of streamwise velocity in the near wall region at the same time.

By the term drag reduction in the present case indicate particularly the viscous drag reduction of the TBL. Bushnell (1983) termed the friction drag reduction as a major "Barrier problem" due to the fact that the further total drag optimization reached its pinnacle for most of the aerodynamic and hydrodynamic bodies. Therefore, selection of effective flow control depends on the optimum amount of viscous drag reduction. In this section, we discuss the effect of MBT on the turbulence process of the TBL.

## 1.7 Drag reduction mechanism

Figure-1.8(a) from Section-1.5, introduced the basic structures present in TBL. What follows in this section is how they interact to each other in terms of the generation of drag.

In the study of structures in TBL in terms of drag reduction, one has to consider various eddy scales, from the smallest scale which is in the size of viscous length scale $(\nu/u_\tau)$ to the largest scale in the order of boundary layer thickness. This is however expressed by the Reynolds number based on shear velocity.

From the wall through the buffer layer and some part of the lower log layer is populated with the longitudinal Quasi Streamwise Vortices (QSV) (also known as Low-Speed-Streaks (LSS)) which are responsible for the violent "Bursting Process (BP)" and hence subjected to the instability which causes lifting up from the wall and "ejection" events (Blackwelder (1989)). They are "wrapped" with strong gradient of shear in transverse and longitudinal directions $(\partial u/\partial z)$ and $(\partial u/\partial y)$.

We now know that near wall turbulent production and maximum velocity fluctuation is caused by such events. Eventually, this leads to the generation of the hairpin like streamwise vortices (SV) (also known as hairpin vortices).

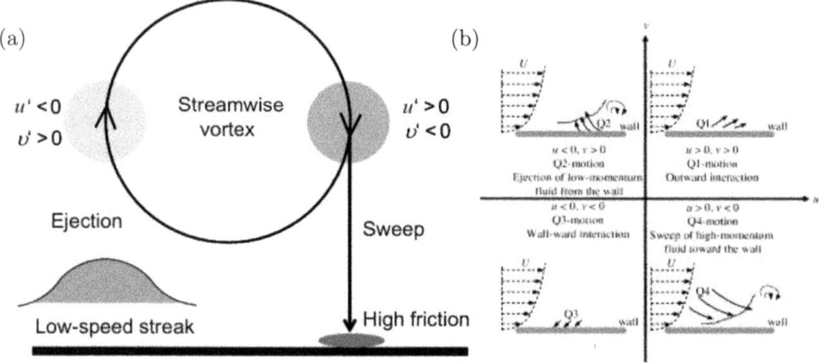


**Figure 1.21:** (a) Schematic diagram of low speed (Xu et al. (2013)) (b) Diagram of the categorization of "ejection" and "sweep" events (Cai et al. (2009)).

Although maximum production, fluctuation and shear stress is contributed within the thin layer of buffer region ($y^+ = 10 \sim 11$) (Fernholz and Finley (1996) and Österlund et al. (1999a)), recent advances show that it is the log region which is responsible for growing influence on the near wall structures at high Reynolds number (Mathis et al. (2009)). Decomposing streamwise velocity measured by single HWA, Amplitude Modulation correlation coefficient was derived (see Hutchins and Marusic (2007a)). This parameter shows the influence of the large-scale motions to the small-scale near wall cycle. Therefore, the structures inhabit the log region, have a significant influence to the drag footprint at the wall.

Adrian (2007) proposed an identification of turbulent structures, where hairpin like structures populates within the boundary of buffer layer outer edge to the log and wake region. Moreover, they form packets of hairpins contained within a large bulge which can also stretch beyond the intermittent outer layer. Individual hairpins are formed with a counter-rotating pair of asymmetric legs, neck and head which is inclined and have streamwise vorticity. The birth and decay of such structures are periodic and has strong correlation to both the "ejection" and "sweep" events. The tangential motion of the vortex core against the wall provides the necessary drive to the low momentum fluid entrained within to move upward and the inward motion of the head directed downstream bring the high momentum fluid towards the wall. Therefore such motions are determined as "sweeps" events and has been proven to be responsible for the high shear stress footprints at wall. How this works is shown in the Figure-1.21 (a) (Xu et al. (2013)).

Corino and Brodkey (1969) determined a second "ejection" event connected to an earlier one and they are also associated to the "sweep" events of the large scale high energy structures which are more linked with the intermittent outer layer. Wallace (2016) proposed the detection method of these basic events connected to the wall bounded flows known as "Quadrant Analysis". Through quadrant analysis we can quantify the Reynolds shear stress contribution from any of these events. How the streamwise and wall-normal velocity components behave during any of these events is shown schematically in Figure-1.21(b)(Cai et al. (2009)). Looking into the turbulence regeneration process, controlling "ejection" and/or "sweep" events can lead towards the desirable outcomes such as drag reduction.

General consensus regarding the QSVs is that they are $20l^{+}$ (where $l^{+}$ is the viscous length scale derived from kinematic viscosity and shear velocity) in diameter and spaced in spanwise direction at $80{\sim}100l^{+}$ distributed as high and low speed streaks (Lee et al. (1974)). On the other hand, streamwise vortices, also denoted as "Hairpin vortices" are formed in a packet where the youngest is the offspring of the oldest one. The inclination angle of such packets in comparison to the wall is ~14.5deg (Marusic and Heuer (2007)). The dynamical relationship between these two sets of structures are however the driving mechanism for the generation of local shear stress. Most importantly, the "Auto-generation"/"regeneration" process of hairpin formation is strongly nonlinear and occurs only at a certain threshold value. Moreover, they form in packets and have re-occurring features with the interval dependent on the Reynolds number. Their growth of scales depends on the distance from the wall and provides a mechanism for the transport process of TKE and low momentum fluid.

Within the context of the present experiment, effect of local blowing through the uniformly perforated surface which injects momentum flux of air at very low velocity ($Re_D<2$), will be studied. We assume that the interaction of the injected air to the incoming flow offers significant changes to the basic "ejection" and "sweep" events controlling the drag footprints at wall. Hasanuzzaman et al. (2020) suggested that the inner layer turbulence is also effected but not as prominently as the outer layer turbulence. Hence, we assume that the MBT disturbs the TBL intensely, causing an amplification of outer layer (spectral) turbulence. MBT creates ordered vortices in the region where LSMs dominate the flow dynamics and subsequently, changing the behaviour of LSM and VLSMs in such a way that "ejection" events are enhanced and "sweep" events are attenuated. Therefore, MBT reduces shear stress at the wall and enhances turbulence in the area where LSM and VLSMs are prominent.

## 1.8 Present experiment

Primary objective of this thesis is to experimentally study the effect of wall normal blowing in a flat plate TBL without a pressure gradient (neglecting ~ very small favourable pressure gradient). Since, Schlichtling (Schlichting (1942a) and Schlichting (1942b)) investigated turbulent boundary layer with injection, a series of subsequent researches has been undertaken which was discussed in the previous sub-sections. Our relative understanding of turbulence modification with flow control compared to the traditional TBL without control is limited. Despite numerous researches on the flow control in wall bounded shear flows, we are yet to find an optimum model to control the turbulent structures due to large spectrum of scales and limitations of the measurement technique. Most of this effort spent on the outcome of such perturbation method in terms of drag reduction as an engineering application.

Therefore, this thesis is a written report of the experimental study in order to investigate statistical and spectral modifications of the flow field in a wide range of Reynolds number spectrum, $Re_{\theta,SBL}$ = 1100 ~ 18000, this is a corresponding value to the inertial condition of $\delta^+_{SBL}$ = 415 ~ 5500. In terms of Reynolds number based on characteristics length of $Re_x$ = 0.38 ~ 13 million. In order to realize this wide ranges of Reynolds number, two experimental facilities were used. The closed return Göttingen type wind tunnel located at the Department of Aerodynamics and Fluid Mechanics is also known as BTU wind tunnel. The other close return wind tunnel from the University of Lille is known as Laboratoire de Mécanique des Fluides de Lille boundary layer wind tunnel, in short LMFL. A detailed description of both these wind tunnel facilities will be discussed in chapter-2 and 4 respectively.

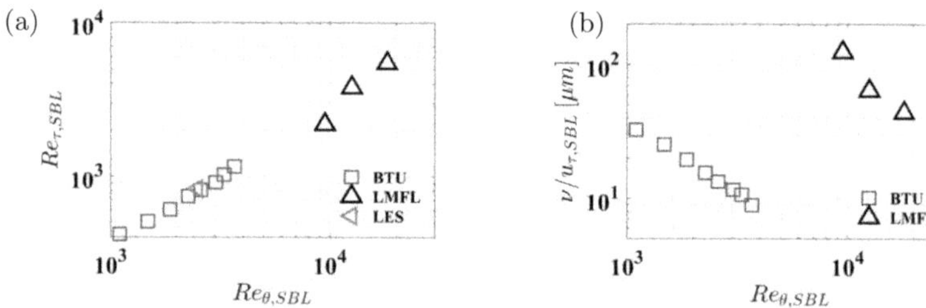


**Figure 1.22:** (a) Present experimental ranges based on their reference smooth wall data for two different wind tunnel facilities. Well resolved LES range is obtained from Kametani et al. (2015).(b) Viscous length scales for different Reynolds number range.

Figure-1.22(a) presents the reference smooth wall Reynolds number measured in logarithmic scale. Here, horizontal and vertical axis presents momentum thickness and shear velocity based Reynolds number respectively. Figure-1.22(b) presents the viscous length scales. However, another challenging task while designing the experiments for two experimental facilities were to take into consideration their viscous length scales and to apply suitable measurement technique in order to obtain reliable data. As seen from this figure, BTU wind tunnel has an operating range of Reynolds number much smaller than that of the LMFL. In addition, due to the smaller plate length of BTU wind tunnel, incoming flow velocity was higher in order to reach the presented Reynolds number range. Which reduces the viscous length scale significantly, therefore, posses greater difficulties to have access and obtain near wall data.

The bounding condition of traditional boundary layer with no slip condition will be replaced with a constant wall normal velocity uniformly distributed over a specific area in the turbulent region. For the subsequent text, this uniform wall normal velocity without any stream-wise component will be termed as "blowing" after Hwang (1997). The aim of this thesis is to study the effect of uniform blowing to the turbulent boundary layer in terms of their morphology, statistics of different Reynolds stresses to quantify the mean turbulent kinetic energy, determination of wall shear under different boundary conditions and quantification of spectral energy of different turbulent structures as an outcome of uniform wall normal blowing.

Measurements from the present thesis is based on the turbulent boundary layer where the downstream pressure distribution is adjusted in such a way that their velocity profiles are independent of the Reynolds number and of the downstream distance from the leading edge of the plate, provided that the profiles are appropriately non-dimensionalized with an *velocity defect law*. Hence, stochastic equilibrium was maintained while preparing the flat plate. Principle velocity component is the longitudinal one, therefore, no change of free stream velocity along longitudinal axis is assumed e.g $\partial U_\infty / \partial x$.

In order to access data presented in Chapter-3, please use the following link and refer to Hasanuzzaman et al. (2020).

`https://turbase.cineca.it/init/routes/#/logging/view_dataset/82/tabmeta`

# 2 Moderate Reynolds number experiment

## 2.1 Experimental setup

### 2.1.1 Wind tunnel

The experiment was realized using a closed-return, Göttingen type, subsonic wind tunnel at Brandenburg University of Technology, Cottbus-Senftenberg, Germany. All the results presented in chis chapter has also been published in Hasanuzzaman et al. (2021). The wind tunnel was designed for good performance upto $U_\infty$=50 m/s, reachable maximum velocity corresponds to the $Re_x = U_\infty L/\nu = 7\times10^6$, where, L is the characteristics length along the flat plate parallel to the principle direction of flow. The contraction ratio and thermal stability are 1:5.5 and ±0.1 K respectively. Thermal stability was maintained using a heat ex-changer assembly installed before the honeycomb grids. As shown in Figure-2.1(a), the wind tunnel has a measurement section with 0.6 × 0.5 $m^2$ cross-section and 1.5 m length. The test section has optical access from top and both the side walls which makes it suitable for non-intrusive measurements. The optical access is made from glass with minimum refractive index deviation as shown in Figure-2.1 (b). While using LDA as a measurement technique, closed loop wind tunnel is more advantageous then an open flow wind tunnel. The flow is undisturbed from indoor flow conditions and other external influences for closed loop flow. Even the flow quality is better due to (corner) turning vanes at corners and screens.

Motion of the particles suspended in air is affected by particle shape, particle size, relative density of particle and air and finally, the concentration of particles in air. This factors were taken into careful consideration while seeding the flow. Therefore, in order to facilitate optical measurements such as LDA, a particle generator (Aterosol generator, Model: AMT 230, Topas Co) was used to produce particles with diameter ≈0.3$\mu$m. Di-Ethyl-Hexyl-Sebacat (DEHS, Topas Co.) was used as the aerosol liquid (dynamic viscosity, $\mu = 0.023$ Pa S, Kinematic viscosity, $\nu = 25.16$ $mm^2$/s, Specific gravity = 914 kg/$m^3$ and vapour pressure < 1 Pa). Continuous motion of the seeding particles in a closed loop was quite appropriate for the present measurement technique (LDA). Most penetration particle size has a lifetime of approximately 4 hours.

The flow pattern in this wind tunnel was uniform and of less turbulence which was slightly greater than 0.5%. A three-dimensional traversing system (Isel Germany AG) was used in order to move the LDA probe as can be seen in Figure-2.1(b) with a minimum step size of 0.0063 mm (less than the size of viscous length scale at maximum Reynolds number).

A total of 32 TBL profiles were measured at constant stream-wise position using LDA. Details regarding measurement technique is explained in Subsection-2.1.3.

### 2.1.2 Flat plate geometry

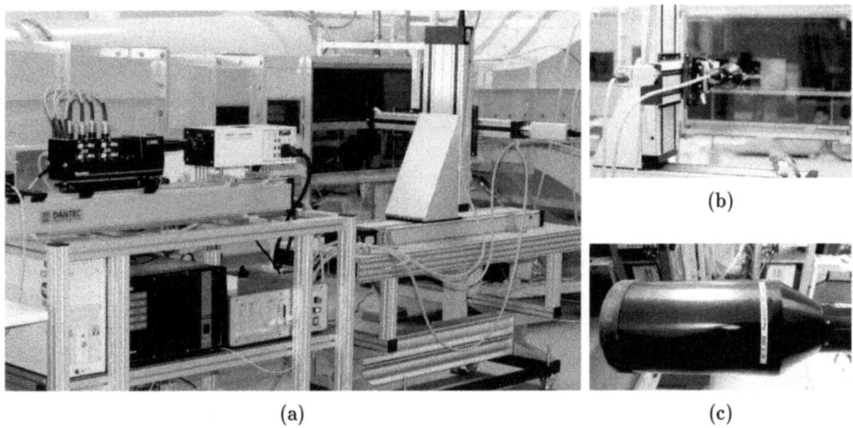

**Figure 2.1:** (a) LDA and traverse setup at LAS wind tunnel, here, principle direction of the incoming flow is from right to left from readers perspective; (b) Laser probe positioning and (c) Optical mounting used for present experiment including beam expander where, effective beam diameter at the front lens is equal to the product of beam spacing and expander ratio.

The flat plate as shown in Figure-2.2 was made from eloxide aluminium with a dimension of $1.055 \times 0.595 \times 0.019m^3$ (length × width × thickness) and was mounted horizontally in the test section as shown in Figure-2.2. The leading edge was made in an elliptical shape followed by Smits et al. (1983) with an ellipse aspect ratio of 4:1[1]. A 'DYMO' brand label printed tape letter 'x' ($0.7 \times 5mm^2$, height × width) along Z axis at a fixed distance from leading edge ≈ 5 % was used as the tripping device for early transition to the turbulent regime (Tripping location is shown in Figure-2.2), see Kito et al. (2006) for more details.

In order to measure the ZPG condition along the flat plate, the plate contains 16 pressure tapping points as can be seen in Figure-2.2. These are located in the center line of the plate along x axis with a distance of 50 mm from each other. However, the first and the last pressure tapping points are 100 mm from the leading and trailing edge of the plate. The pressure tapping points have a diameter of 0.5 mm. Each opening is

[1]ellipse aspect ratio is the ratio between major axis along X axis to the minor axis along Y axis

fitted with an approximately 20 mm long metal tube with a diameter of 1.9 mm from the bottom side of the plate.

The pressure difference between the two different measuring positions on the plate was measured using a pressure measuring device GMH3155 (Greisinger electronic GmbH). The device has a piezo-resistive relative pressure sensor GMSD 25MR with a resolution of ±0.01 mbar. The pressure gradient $\Delta P$ was measured according to the scheme described by reference Klebanoff (1954). The pressure of the last opening served as a point of reference. Last tapping point is considered as the reference pressure point $P_0$. At the end, the pressure gradient $\Delta P = (P - P_0)/q_0$ where $q_0$ is the dynamic pressure, which was determined by a pitot tube located in the free stream directly above the last tapping point. As can be seen from Figure-2.3, fairly ZPG zone is observed at $x/L = 0.4 \sim 0.7$.

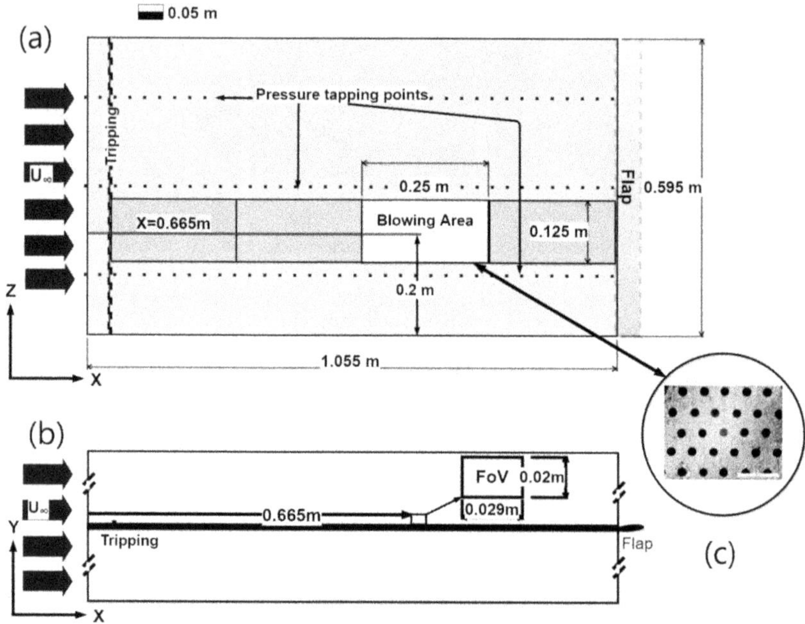


**Figure 2.2:** Schematic of flat plate used. In order to describe the geometry, origin of the Cartesian co-ordinate is selected on the bottom-left corner of the plate. Therefore, streamwise direction parallel to the principal flow is indicated with X, positive spanwise direction from bottom-left corner is indicated with Z and positive distance away from the wall is indicated with Y.

Immediately after the tripping device, a segment of the flat plate was installed together with 4 different pieces of small sections. This way of construction allows us to change the blowing area selection in different x directions. For present experiment, two smaller sections of smooth surfaces ($250 \times 125 \times 14 mm^3$, length × width × height) were prepared to install. Subsequently, perforated surface section was installed in the third slot. With the plates present arrangement, length of the perforated surface started at ≈52% and ends at ≈70%. Blowing area is indicated as white area in Figure-2.2(a).

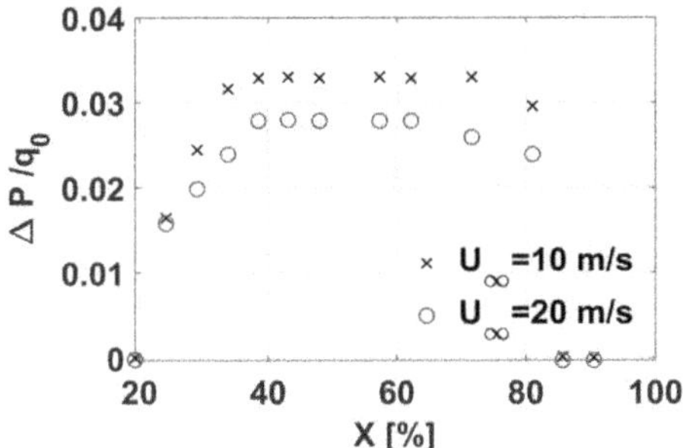


**Figure 2.3:** Pressure distribution along the flat plate.

Reference SBL measurements were taken installing smooth anodized aluminum plate to obtain a non shiny black surface in order to avoid LDV light reflection. To apply constant blowing, a second plate of 1 mm thickness ($t_h$) was manufactured with the same exterior dimension (125×250 mm$^2$, width×length) with electron beam drilled holes, uniformly distributed over the entire surface. Figure-2.2 (c) represents the microscopic view of the perforated region. Each hole in the surface having a diameter ($d_h$) of 0.18 mm ($20 \times \nu / u_\tau$ at maximum Re no.) and equidistant to each other by 0.4 mm. Porosity of the perforated plate ($p_h$) is 18 % and aspect ratio ($ar_h = t_h/d_h$) is 5.55, a larger $ar_h$ allowed us to provide a laminar flow of blowing even at higher flow velocity as Reynolds number of each holes is very low ($R_D = U_b d_h / \nu \leq 2$ based on hole diameter ($h_d$) and blowing velocity ($U_b$) as characteristics length). The dimension of the blowing area inside the pressurized chamber of the blowing assembly is 0.23 × 0.105 m$^2$ in length ($L_i$)× width ($w_i$). Therefore, exit area of the perforated surface $S_f = L_i \times W_i \times p_h$ is obtained.

Porous surface is attached to a pressurized chamber with multiple layers of screening surfaces in order to distribute the air pressure evenly inside the chamber which can be seen in Figure-2.4(a). The design of the present perforated surface is adopted from Motuz (2014), where staggered distribution[2] was followed to implement the micro-perforated holes.

[2]Very recently, Horn et al. (2015) suggested that staggered holes arrangement is as effective as discrete perforations. Therefore, Tailored Skin Single Duct (TSSD) design was implemented for Airbus A340-300 with a hole diameter to spanwise distance ratio of 1:2. Scale modification was done based on $\delta$, this was discussed in detail from Krishnan et al. (2017).

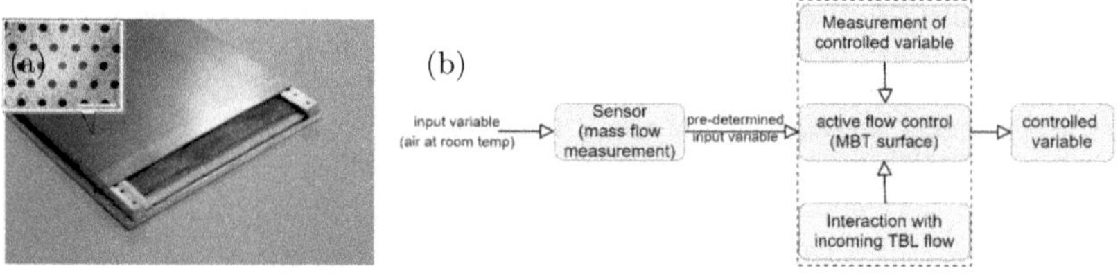


**Figure 2.4:** (a) Different layers of MBT assembly, (b) MBT process diagram.

Air (at the same temperature as the wind tunnel air) is supplied from a air compressor through a digital flow meter, DFM-47 (Aalborg 47), with an accuracy of $\pm 1$ % via hermetically sealed pneumatic system. Using the flow meter, flow rate ($Q_c\ [m^3/s]$) is measured, and blowing velocity, $U_b = Q_c/S_f\ [m/s]$ is calculated analytically using control volume analysis using porous area, porosity and flow rate. Afterwards, analytically obtained $U_b$ is verified using LDA data taken at a wall normal distance of 0.013 mm. The rate of blowing is expressed as blowing ratio defined earlier. Figure-2.4 exhibit the experimental schematic as a process flow diagram. Here, blowing air with a pre-determined input variable as air is injected to the TBL flow, measurement is done using LDA and friction coefficient ($C_f = 2(u_\tau/U_\infty)^2$) is obtained as controlled output. Although, this experiment was not designed based on interactive flow control strategies rather based on pre-determined fixed input based flow control strategy. However, change in turbulent statistics and integral properties are focused at present.

### 2.1.3 Laser Doppler Anemometry (LDA)

Measurements were performed at x = 58 % in z = 33 % which represent the stream-wise and span-wise position respectively over the plate (Figure-2.2) with respect to the origin. Accurate measurement of wall shear is paramount to determine the viscous scaling of the flow. On the other hand, emphasis was given to precise measurement of the wall normal distance (y). However, there are several method to measure the wall shear and friction co-efficient ($\tau_w$ and $C_f$), according to reference Smits et al. (1983), 2D measurement of the velocity component near the wall where $y^+ = yu_\tau/\nu < 10$ is suitable for flow within moderate Reynolds number.

A part of the present results are from measurements using the Laser Doppler Anemometry method. The advantages of using LDA system are Non-intrusive optical measurement and typically, no calibration is often required. In addition, there is well-defined directional response with high spatial and temporal resolution. As discussed in Chapter-1, LDA offers good spatial resolution particularly, suitable for near wall measurements for moderate Reynolds number TBL with high velocity. This technique also offers multi-component measurements e.g one can simultaneously measure different velocity components.

Working principle of the LDA and different variants of its technical realization have been described in detail by several authors, Interested readers are suggested Durst et al (1996), Albrecht et al. (2003), Eder et al. (2012) and Dantec Dynamics (2014) for detailed information on the theory of LDA.

**Figure 2.5:** LDA measurement arrangement at the test section of LAS wind tunnel.

Figure-2.5 exhibit the segment of the BTU wind tunnel where LDA setup is placed beside the wind tunnel test section. Here, LDA probe is connected with a three dimensional traverse system. Details of different components will be described in later sections.

### Working principle

The measurements presented in this chapter were realized by means of an LDA measurement system in the backward scattering arrangement (integrated transmitting and receiving optics in a common housing) of the Dantec Dynamics Co., see Figure-2.5. An Argon-ion continuous wave laser (Ion laser technology, Salt-lake city, Utah) was used as the laser source. Spatial and temporal coherence of the Argon-ion gas laser make it well suited for the measurement of turbulent flow properties. At all cross sections along the laser beam (in all three dimensions), the intensity has a Gaussian distribution, and the width of the beam is usually defined by the edge-intensity being $1/e^2 = 13\%$ [3] of the core-intensity. At one point the cross section attains its smallest value, and the laser beam is uniquely described by the size and position of this so-called beam waist.

Figure-2.6 exhibit the LDA laser which is a monochromatic beam propagating in z direction (here, minimum beam waist is assumed to be at the origin and positive and

[3] The $1/e^2$ width is equal to the distance between the two points on the marginal distribution that are $1/e^2 = 0.135$ times the maximum value

negative deviation along z indicate distance from the beam waist) with the wavelength ($\lambda$). The beam divergence ($\alpha$) is indicated with Equation-2.1 and is however, smaller than indicated in Figure-2.6.

The laser beam appear to be straight with a constant thickness with mere eyes. The beam radius (where, beam radius being $2r_0 = d_0$) varies along the propagation direction according to Equation-2.1. The beam diameter at wave front (e.g $d_z$, where, z = z in Figure-2.6) is given by Equation-2.3.

Wave front radius R(z) is given by Equation-2.4 which approaches $\infty$ for z approaching 0 e.g the wave fronts are approximately in the immediate vicinity of the beam waist. This behaviour can be interpreted as the theory of plane waves and can be used here which simplifies the calculation of seeding velocity.

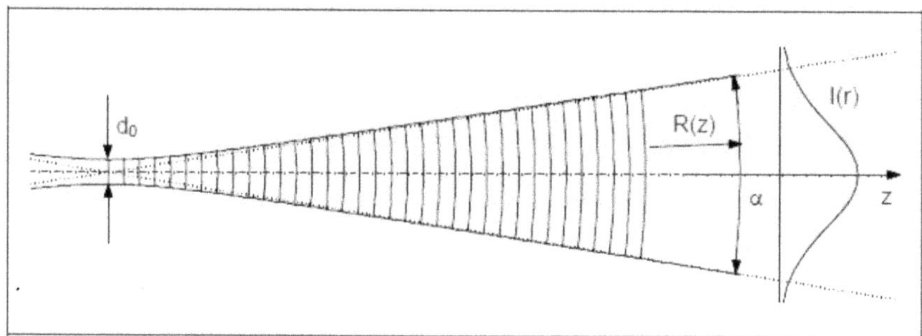


**Figure 2.6:** Laser beam with Gaussian intensity distribution (Dantec Dynamics (2006))

$$\text{Beam divergence,} \quad \alpha = \frac{4}{d_R} = \frac{4\lambda}{\pi d_0} \tag{2.1}$$

$$\text{Rayleigh length,} \quad d_R = \frac{\pi d_0}{\lambda} \tag{2.2}$$

$$\text{Beam diameter,} \quad d_z = d_0\sqrt{\left(1 + \frac{4\lambda z}{\pi d_0^2}\right)} \quad \text{for} \quad \alpha z \quad \rightarrow \alpha \tag{2.3}$$

$$\text{Wave front radius,} \quad R(z) = z\left[1 + \left(\frac{\pi d_0^2}{4\lambda z}\right)^2\right] \tag{2.4}$$

since LDA is based on Doppler shift principle of the light reflected/refracted from a moving seeding particle, Figure-2.7 (a) describe this principle where, we assume that the $\vec{U}$ represent the velocity of a seeding particle in the flow field, and the unit vectors $\vec{e_1}$ and $\vec{e_2}$ describe the direction of incident light ($\vec{e_i}$) of the shifted and un-shifted laser beams respectively, $\vec{e_s}$ gives the direction of scattered light. $\rightarrow$ will not be used for subsequent description of the Laser Doppler Anemometry (LDA) technique.

According to Lorenz-Mie scattering theory, the light is scattered in all direction at once but as shown in Figure-2.7 (a), we consider only the light reflected in the direction of receiver. As such, both incoming laser beams are scattered towards receiving optics, but with slightly different frequency due to the different angle of incidence.

Let us assume, frequency of the scattered light $e_1$ be $f_1$ and $e_2$ be $f_2$. The frequency of the light reaching the receiver can be calculated using Doppler theory. Considering two scattered light received be $f_{S,1}$ and $f_{S,2}$ which is obtained using Equation-2.5. Here, U and c express the particle velocity and velocity of light respectively, therefore U/c ■ 1 even for supersonic flows which means that LDA can be employed to measure very high velocity flows.

The beat frequency corresponds to the differences between the 2 scattered light received e.g $f_{S,1}$ and $f_{S,2}$. As stated earlier, both the incoming laser originates from the same laser source and have same frequency ($f_I$) e.g $f_1 = f_2 = f_I$, where subscript I refer to the incident light from laser source as seen from Figure-2.8. Equation-2.6 gives us the beat frequency.

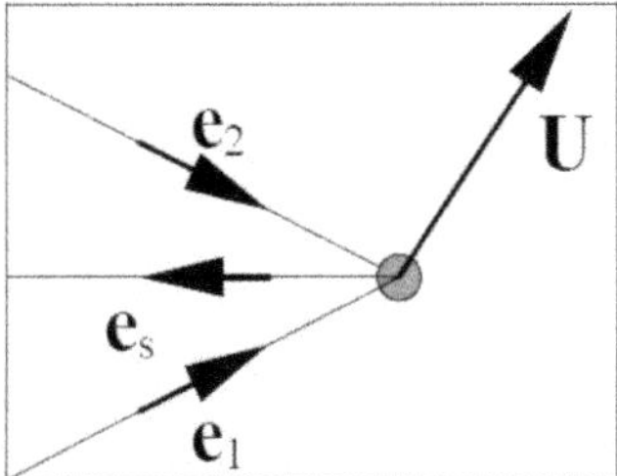


**Figure 2.7:** Scattering of two incoming laser beams for a seeding particle (Dantec Dynamics (2006)).

$$f_{S,1} = f_1\left[1 + \frac{U}{c}(e_s - e_1)\right]$$
$$f_{S,2} = f_2\left[1 + \frac{U}{c}(e_s - e_2)\right] \tag{2.5}$$

Replacing $f_{S,1}$ and $f_{S,2}$ in Equation-2.6, we can obtain the relationship between the beat frequency and the velocity of the particle. Here, $\theta$ is the angle between the incident laser beams.

$$\text{beat frequency,} \quad f_D = f_{S,2} - f_{S,1} = \frac{2sin(\theta/2)}{\lambda} u_x \tag{2.6}$$

or,

$$u_x = \frac{\lambda}{2sin(\theta/2)} f_D \tag{2.7}$$

As it can be seen from Equation-2.7, direction of scattered light ($e_s$) is no longer relevant, this can be interpreted as the position of the receiver has no direct influence on the frequency of the received light measured.

In principle, Beat frequency or Doppler frequency ($f_D$) is much lower than the frequency of the light itself, therefore, it can be measured as fluctuations in the intensity of the light reflected from the seeding particle. For a fixed wavelength ($\lambda$) at a fixed angle ($\theta$) of incoming laser beam, Equation-2.7 indicate that the x-component of particle velocity is directly proportional to corresponding doppler or beat frequency e.g $u_x \propto f_D$. This is to say that particle velocity ($u_x$) can be directly measured from the doppler or beat frequency ($f_D$).

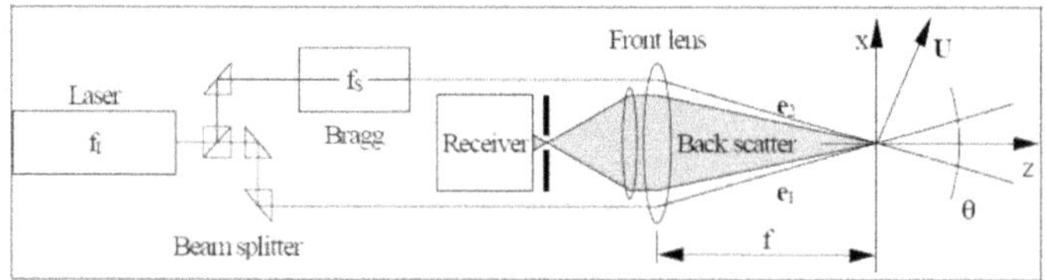


**Figure 2.8:** Optical set-up of a dual beam differential or fringe LDA system in back scattering mode (Dantec Dynamics (2006)).

In Figure-2.8, shows the optical arrangement of the present LDA system in back scattering mode. As shown, the transmitting optics which includes beam splitter which is a dual-beam differential system, where a coherent laser beam of wavelength ($\lambda$) is split equally. One beam passes through a brag cell [4] where frequency of one beam is decreased by the shift frequency $f_s$. Different optical paths of shifted and unshifted beam is compensated through a cylindrical lens by allowing the unshifted beam to pass through it. Finally, a monochromatic lens [5] installed within the transmitting optics focuses both the splitted beams at same focal length using a beam expander so that they intersect with the angle $\theta$. Here, $\theta$ determines the sensitivity as well as the measured velocity

[4]Brag cell is an acousto-optical modulator and requires a signal generator which is typically operated at 40 MHz

[5]Monochromatic lens is a lens designed to limit the effects of chromatic and spherical abberation. Typically used to bring 2 beams of different wavelength into focus in the same plane.

range of the system. The angle $\theta$ can be altered by using a front lens with a different focal length.

For present set-up, focal length of the front lens from transmitting optics ($f_D$) is $400 \pm 0.5\%$ mm. This is also used for the calculation[6] of the fringe spacing and the dimension of measurement volume.

### Calculation of fringe volume

According to Albrecht et al. (2003), accuracy and detection of the measurement volume size depend on the following factors, such as transmission optics (laser power, optical losses, etc.), geometry of the measurement volume (optical configuration), refraction index of the medium, scattering properties (particle properties, observation direction), receiving optics (aperture size, optical loss, etc.), sensitivity of the received signal, signal-to-noise ratio (SNR) and finally, signal detection and signal processing.

On the other hand, the spatial resolution is essentially determined by the detection of fringe volume size ($V_d$), which depends on the following variables; Diameter of the transmitted beam waist and their half angle of intersection as defined as $d_0 = 2r_0$ and $\theta/2$ respectively, spectral intensity[7], spatial filters (in order to overcome imperfections in the receiving optics), sensitivity of the receiving optics.

Two coherent laser beams for the same component intersect at the measurement volume. When they intersect each other, the wave fronts at their beam waists is of approximately uniform in shape. At their intersection they form an ellipsoid as shown in Figure-2.7(a). It is therefore, due to the Gaussian intensity distribution of the laser beams, the measuring volume looks like an ellipsoid. Depending on their direction of incident, largest axis of the ellipsoid is formed. As there are one pair of laser beams for each component of velocity and 2D LDA system was used for the present experiment, two pair of laser beams were applied where each pair have different wavelengths. However, in order to describe the principle, only a pair of beams for the horizontal velocity component will be discussed here for simplicity.

Subsequently, interference of the laser beams produce parallel planes of bright and dark region within the measurement volume also known as fringe. The schematic of fringes can be seen from Figure-2.7(b). Fringes are distributed in parallel along the z axis and their distance from each other is defined as $d_f$. Number of fringes in this measurement volume is obtained by Equation-2.8 which is a relationship between the beam waist diameter ($d_0$), the laser beam wavelength ($\lambda$) and half angle of laser beam intersection ($\theta$). In reality, $N_f$ is effected by the particle residence time within the measuring volume[8] ($\Delta t$) and the beat frequency (see previous section), therefore, obtained as $N_f = f_D \Delta t$.

However, beam waist diameter ($d_0$) can be obtained with the help of Equation-2.9,

[6] by definition, effective beam diameter at the front lens is equal to the product of beam spacing and expander ratio e.g resulting beam diameter ($d_f$) = beam diameter ($d_z$) × expander ratio at the exit of the beam expander or entry to front lens

[7] larger particles with higher intensity of incident light and scattering angle with higher light intensity result in a larger detection volume at a constant detection threshold

[8] time required to cross the measuring volume to cross $d_x$

where, $f$, $E$ and $d_I$ are the focal length of the front lens, beam expander ratio and beam waist diameter before front lens respectively.

$$\text{Number of fringes, } N_f = \frac{d_x}{d_f} = \frac{\frac{d_f}{cos(\theta/2)}}{\frac{\lambda}{2sin(\theta/2)}} = \frac{2d_0}{\lambda}tan(\theta/2) \tag{2.8}$$

$$\text{beam waist diameter, } d_0 = 2r_0 = \frac{4f\lambda}{\pi E d_I} \tag{2.9}$$

The measurement volume depends exclusively on the transmitting optics and the resulting beam geometry. In particular, laser beam waist diameter ($d_0$), beam-half angle ($\theta$) and distance of the beam waists from the transmitting optics (at the focal length) ($f$).

$$d_x = \frac{d_0}{cos(\theta/2)}; \qquad dy = d_0 = 2r_0; \qquad d_z = \frac{d_0}{sin(\theta/2)} \tag{2.10}$$

When a particle move across this fringe region, this produces a burst signal. A total burst signal consists of constant and alternating signals, this can be divided with the help of a high-pass filter into a DC component (doppler pedestal) and an AC component (modulation component) with the aid of a high-pass filter. The envelope of the Doppler pedestal reflects the Gaussian intensity distribution in the measuring volume.

The measurement volume size is the range in which the AC power amplitude $I_{AC}$ is greater than $e^{-2}$ which is relative to the maximum AC power amplitude at the position [0,0,0] $I_{AC,max}$. It is also independent of the scattered particle data. In summary, only the AC power signal is called a laser Doppler signal.

Detection of the fringe volume ($V_d$) is defined by the relationship between the maximum amplitude of alternating component ($I_{AC,max}$) of the burst signal, since this is the only parameter that is used for the velocity determination of the particle. This means that during an LDA measurement only signals that exceed a defined intensity threshold ($I_d$) are detected. Therefore, selection threshold for default anode current was set to be -2 and 0 dB for streamwise and wall normal component respectively. However, selection of higher values (optimized) will reject more noisy bursts leading to better validation results.

Since the intensity of the light scattered by a particle depends on its size, the fringe volume is also dependent on the particle size. The fringe volume $V_d$ is calculated as Equation-2.11.

$$\text{fringe volume, } V_d = V_0\left(\sqrt{\frac{1}{2}ln\frac{I_{AC,max}}{I_d}}\right)^3 \tag{2.11}$$

$$\text{measurement volume, } V_0 = \frac{8\pi r_1^3}{3sin\theta} \tag{2.12}$$

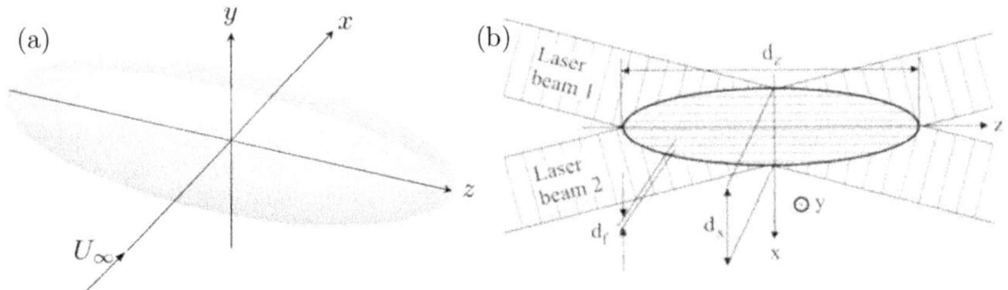


**Figure 2.9:** (a) Measurement volume of a laser Doppler anemometer and (b) Fringe distribution in xz plane (Eder et al. (2012)).

| Parameter | U-Component | V-Component |
|---|---|---|
| Wave length, ($\lambda$) [nm] | 488.0 | 514.5 |
| Focal length (f), [mm] | 310.0 | 310.0 |
| Beam diameter ($d_0$), [mm] | 1.35 | 1.35 |
| Extension ratio of the beam expander (E) | 1.98 | 1.98 |
| Beam spacing, [mm] | 38.0 | 38.0 |
| Frequency shift of bragg cell, [MHz] | 40 | 40 |
| SNR level, [dB] | -2 | 0 |
| Number of fringes ($N_f$) | 35 | 35 |
| Fringe spacing ($d_f$), [$\mu$ m] | 2.025 | 4.205 |
| Beam-half angle ($\theta/2$) [deg] | 6.919 | 3.507 |
| Measurement length in x - dx [mm] | 0.07259 | 0.1507 |
| Measurement length in y - dy [mm] | 0.07206 | 0.1504 |
| Measurement length in z - dz [mm] | 0.5982 | 2.459 |
| Minimum increment of the traversing system, ($\mu m$) | 6.35 | |

| $Re_\theta$ | 1100 | 1480 | 1870 | 2270 | 2590 | 3030 | 3300 | 3670 |
|---|---|---|---|---|---|---|---|---|
| $dx^+$ | 2.2 | 2.8 | 3.7 | 4.6 | 5.4 | 6.2 | 6.8 | 8.2 |
| $dy^+$ | 2.2 | 2.8 | 3.7 | 4.6 | 5.4 | 6.2 | 6.8 | 8.1 |
| $dz^+$ | 18.3 | 23.6 | 30.6 | 38.4 | 44.6 | 51.4 | 56.1 | 67.2 |

**Table 2.1:** Dantec Dynamics LDA parameters used for present measurement system

This is much smaller than the measurement volume ($V_0$) which is expressed with the help of Equation-2.12. The 3D traversing system from Isel used for positioning the LDA probe has a minimum increment of 6.35 $\mu m$. A relatively small measurement volume, along with a precise traverse system, provides very good spatial resolution in order to determine the flow velocity and allows the measurement in a very small area near the wall. Direct measurements from near wall region therefore, particularly important in turbulent boundary layer studies.

Table-2.1 present the properties of the 2D LDA system, some important parameters of the measuring system used in the experiment are shown. The size of the measuring volume for streamwise velocity component was $0.07259 \times 0.07206 \times 0.5982$ mm$^3$. This volume corresponds to the $2.25 \times 2.25 \times 18$ times the viscous length scale of the minimum Reynolds number obtained for the present experiment, $8 \times 8 \times 66$ for maximum Reynolds number. The dimensional thickness of there viscous sub-layer of the boundary layer under investigation was in the range of $165 \sim 45\mu m$ ($\sim 5y^+$) depending on the Reynolds number. The longest axis of the measurement volume was aligned parallel to the wall surface along z axis e.g 8 times the size of length in x or y direction of the probe volume. Therefore, more particles are detected and measured in spanwise direction which is the pre-requisite to the Prandlt's 2D assumption of the boundary layer profile in the viscous sub-layer (Prandlt (1927)).

Spatial resolution of the probe dimension effects the measurement strongly, this is even stronger at higher Reynolds number measurements. This spatial resolution can provide false information of the RMS values of the velocity data. Moreover, false estimation of the small scale contribution to the turbulent fluctuation can cause failure to the outer peak analysis and as the Reynolds number increases, increased probe volume will sense only the large scale contribution to the velocity signal (Hutchins et al. (2009)). Dimensions of probe volume in viscous-scales units is given in Table-2.1, probe length for x and y component varied between $2 \sim 10\, y^+$ as suggested by Ching et al. (1994). Special precaution was taken in order to keep the non-dimensional probe length along x and y below 10 as suggested by Hutchins et al. (2009). On the other hand, z-wise length was kept larger in order to have more averaging in spanwise direction.

## 2.2 Processing

### 2.2.1 Data acquisition and processing

Enhanced burst spectrum analyzers (BSA, Dantec Dynamics) were used to process the photo multiplier signals. In order to obtain bias free convergence of the data, minimum number of total samples for u component started at 1000 and reached upto a maximum of 10000, for v-component upto a maximum of 5000 samples were recorded. Maximum time of acquisition was set to 60 seconds which corresponds to $t^+ = tu_\tau^2/\nu = 8.85e+10 \sim 1.2e+7$ for all the Reynolds number investigated. Balakumar et al. (2007) suggested that the largest scales present in the TBL could be comparable to the size of $20\delta$. Therefore, total sampling time (T) should be large enough for good statistical convergence. This can be measured with boundary layer turn over times and calculated as $T^+ = T/(\delta/U_\infty)$, for present experiment which varied between $T^+ \approx 4.5e+3 \sim 2.5e+5$. Number of validated bursts per second on a logarithmic scale which is also known as data rate[9] was within

[9] total number of samples/($\text{end}_{time}$ - $\text{start}_{time}$)

1000 to 11000. Data validation rate was taken into consideration and was kept greater than 50%. This was obtained for each data points as a percentage of valid bursts/(valid bursts + invalid bursts)).

Statistical moments were calculated using the transit time weighing following Equation-2.13, where, N is the number of total samples and $t_i$ is the transit time of the $i^{th}$ seeding particle passing the LDA measuring volume. Subequently, mean, root-mean-square, skewness and flatness of u-component of velocity was calculated using Equation-2.14, 2.15, 2.16 and 2.17 respectively. Similarly moments of v-component was also calculated as u component.

$$N = \frac{1}{\eta_i} = \frac{\sum_{j=0}^{N-1}(t_j)}{t_i} \tag{2.13}$$

$$\bar{u} = \sum_{i=0}^{N-1} \eta_i u_i \tag{2.14}$$

$$u_{rms} = \sqrt{\sum_{i=0}^{N-1} [\eta_i (u_i - \bar{u})^2]} \tag{2.15}$$

$$S(u) = \frac{1}{u_{rms}^3} \sum_{i=0}^{N-1} [\eta_i (u_i - \bar{u})^3] \tag{2.16}$$

$$F(u) = \frac{1}{u_{rms}^4} \sum_{i=0}^{N-1} [\eta_i (u_i - \bar{u})^4] \tag{2.17}$$

### 2.2.2 Wall shear stress

Scaling on wall variables e.g friction velocity ($u_\tau$) and kinematic viscosity ($\nu$) is a common way of scaling of a TBL. Usually this is done based on Clauser fitting (Clauser (1956)) to the logarithmic law (Equation-1.18, according to Österlund et al. (1999a), constants $\kappa = 0.38$ and $C = 4.1$). However, at low Reynolds number this fitting becomes quite difficult due to very small logarithmic region. Although at moderate Reynolds number experiments such as the present one, fitting based on inner scaling is no longer possible when wall normal blowing is applied as it changes the friction velocity. On the other hand, inner length scale decreases rapidly as the Reynolds number grow. In addition, direct measurements from oil film interferometry (OFI), HWA or preston tubes measurements may be inaccurate as described in Winter (1977).

Assuming 2D boundary layer in near-wall region which was confirmed by keeping the larger side of the LDA probe aligned in z-direction, slope of the mean velocity profile was detected using a sliding regression method. The sliding regression detected the correlation co-efficient (CorrCoeff($\bar{u}$)) greater than a threshold value ($\geq$0.995), therefore, linear region was detected for the corresponding data points with uncorrected wall distance.

$$\text{CorrCoeff} = \frac{1}{1-N}\left(\frac{\bar{u}_i - \mu_u}{\sigma_u}\right)\left(\frac{ydata_i - \mu_{ydata}}{\sigma_{ydata}}\right) \qquad where,\ N = 2, 3, 4, ....12 \tag{2.18}$$

In Figure-2.10(a), detection of data points in linear region with CorrCoeff $(\bar{u}, ydata)$ above 0.995 is shown at $Re_{\theta,SBL}$ = 1480 along there uncorrected wall distance (ydata). Here, number of samples varied between 2~12 for sliding regression of linearity (Equation-2.18). Finally, a stable linear plateau of CorrCoeff$(\bar{u}, ydata)$ values were detected for an average deviation of the mean value within ≤ 1%. Beginning of the plateau (indicated as red x in Figure-2.10(a)) provides the first valid data point of the profile. Once the data is detected within the prescribed threshold values (0.995 < CorrCoeff < 1), number of uncorrected data points within this threshold is indicated as y. Thus, a linear fit of the relationship $\nu(d\bar{u}/dy)_{y=0}$ using y provides the wall distance correction $(y_0)$. Finally, the slope of the corrected du/dy is used to obtain wall shear $(\tau_w)$ as shown in Figure-2.10(b). Subsequently, friction velocity $(u_\tau)$ and friction co-efficient $(C_f)$ is calculated for all data sets.

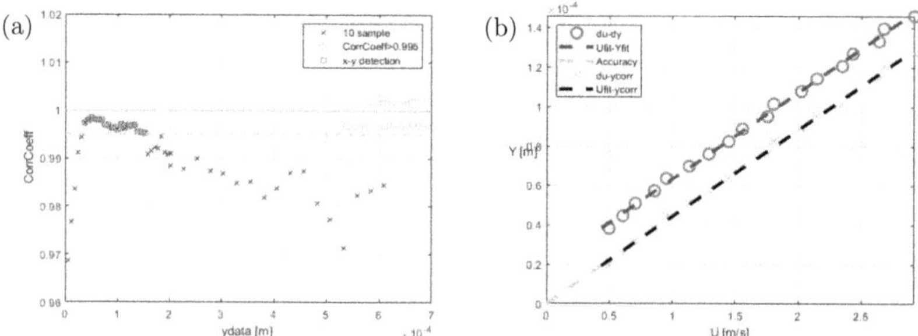


**Figure 2.10:** (a) Detection of the linear near wall region using the plateau of CorrCoeff at $Re_{\theta,SBL}$ = 1480, here 'x'; indicates CorrCoeff obtained with 10 samples, '○'; CorrCoeff ≥0.995 and '□'; are the detected 'x' and 'y' value. (b) '○'; detected uncorrected data points in linear near wall region, '×'; data points after wall distance correction.

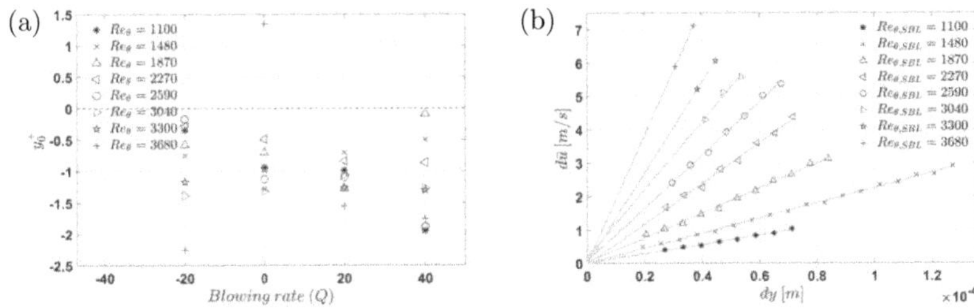


**Figure 2.11:** (a) Estimated wall distance correction applied to the near wall data. (b) Linear velocity gradient for different Reynolds number in dimensional form.

Figure-2.11(a) presents the corrected wall distance normalized with inner scaling ($y_0^+$) for all the measurements presented in this chapter. Wall distance correction was limited to $y_0^+ \leq 4$. Figure-2.11 (b) presents the streamwise velocity gradient along wall in dimensional form. The gray continuous line is the fitted line through linear regression. The figure also indicates that the number of data points gradually decreases as the Reynolds number increases, which is a consequence of smaller viscous sub-layer.

### 2.2.3 Error analysis

According to Durst et al (1996), statistical uncertainties and measuring errors are the source of error in experimental data. Therefore, they advised to reduce the statistical uncertainties and quantify the error limit. In principle, measurement errors are classified as systematic and random error. McLaughlin and Tiederman (1973) has discussed the systematic error in details but to quantify the systematic error (such as optical access and surface interference) is quite difficult for LDA measurement system.

$$U_{\bar{u}} = \pm 1.96 \times \sqrt{\frac{u_{rms}}{N_{eff}}}$$
$$U_{u_{rms}} = \pm 1.96 \times \sqrt{\frac{u_{rms}}{2N_{eff}}} \tag{2.19}$$

On the other hand, one of the most significant source of non-random error for LDA system is the use of finite size of the probe volume (also known as fringe volume or measurement size). The size of the particles used to seed the flow is substantially smaller than the vertical size of the fringe volume. For measurements over a sufficiently long period of time, it is possible to detect more faster particles than the slower ones passing through this window. As a result, time average velocity measurement can lead to over estimation of the mean velocity. In order to avoid this over estimation, McLaughlin and Tiederman (1973) suggested an improved estimation of the mean velocity using weighted average (Equation-2.13) of each velocity realizations provided that the LDA system is operated in continuous wave mode (see subsection-2.1.3, present LDA system is operated in continuous wave mode). Thus, Equation-2.14 provides better estimation of the mean velocity and can help reducing the statistical biasing present in the data.

| $Re_\theta$ | $U_{\bar{u}}$ | $U_{u_{rms}}$ | $U_{\bar{v}}$ | $U_{v_{rms}}$ |
|---|---|---|---|---|
| | $[\%U_\infty]$ | $[\%U_\infty]$ | $[\%U_\infty]$ | $[\%U_\infty]$ |
| 1100 | 0.22 | 0.15 | 0.15 | 0.10 |
| 1480 | 0.27 | 0.19 | 0.21 | 0.15 |
| 1870 | 0.25 | 0.17 | 0.19 | 0.14 |
| 2270 | 0.27 | 0.19 | 0.21 | 0.15 |
| 2590 | 0.25 | 0.18 | 0.20 | 0.14 |
| 3030 | 0.28 | 0.20 | 0.23 | 0.16 |
| 3300 | 0.34 | 0.24 | 0.27 | 0.19 |
| 3670 | 0.48 | 0.34 | 0.36 | 0.26 |

**Table 2.2:** Uncertainty estimation of the mean and rms calculation of the u and v-component data from LDA presented as % of $U_\infty$ e.g ($\_/U_\infty \times 100$). 1% rounding off of the data is used for clarity.

Statistical biasing increases in the region where turbulence intensity is high. Although the histogram of velocity realizations are biased to the higher values, a good estimation of Root-Mean-Square (RMS) can also be obtained from the biased ensembles using Equation-2.15.

According to Benedict and Gould (1996), assessment of statistical uncertainty for random sampling measurements for LDA system is given as Equation-2.19, where $U_{\bar{u}}$ and $U_{u_{rms}}$ present the uncertainty estimation of mean ($\bar{u}$) and rms ($u_{rms}$) calculation of u component respectively with 95% confidence interval. Similarly, uncertainty of mean and rms calculation of v-component was also carried out and presented as a percentile of corresponding $U_\infty$ in Table-2.2. This results belongs to the reference smooth wall measurements and within less than 0.5 % of $U_\infty$.

The error associated to the wall shear determination will be discussed in the rest of this section. According to Durst et al (1996), this is divided into four parts. First step is to obtain sufficient number of particles in the near-wall region providing enough valid data through burst signal analyzer. Next is to validate the random error and statistical biasing of the experimental data, uncertainty in determining absolute wall distance and finally, the residual error associated to the linear regression approach.

In order to obtain sufficient number of data points, special seeding was arranged from the leading edge section of the plate. Eventually, measurements were repeated a few times at the same Reynolds number in order to confirm sufficient number of data points for near wall region. However, minimum number of statistically independent streamwise velocity samples e.g $N_{eff} \leq 2000$ was achieved.

As discussed earlier, statistical biasing is even more prominent due to the finite size of probe volume in the region of high shear (and high velocity gradient). Tiederman and Reischman (1975) suggested that using weighted mean (Equation-2.14) is enough to avoid biasing in the viscous sub-layer as long as the velocity gradient is linear. Therefore, no additional correction was made in the data from near wall region to avoid statistical bias.

Assuming the errors from the mean velocity measurement following a normal distribution, near wall data presents RMS error $\epsilon/\bar{u} \le 1.9\%$ based on the relationship from Lumley (1970) which is expressed as Equation-2.20.

$$\frac{\epsilon}{\bar{u}} = \frac{\sqrt{2}}{\sqrt{N_{eff}}} \frac{u_{rms}}{\bar{u}} \tag{2.20}$$

Getting wall shear stress from direct velocity measurements (at $y^+ \le 5$) suffers from the uncertainty in determining absolute distance from the wall. As discussed in the Durst et al (1996), uncertainty of determining corrected wall distance ($y_0$) is almost the half of vertical height of the probe volume. For present case, the probe volume was the same for all the Reynolds number measured. Therefore, uncertainty for accurate wall distance ($\epsilon_{y_0}$) was $1.1y^+ \le \epsilon_{y_0} \le 4y^+$ depending on the corresponding Reynolds number. As discussed in the previous section in Figure-2.11(a), uncertainty in the wall correction is very small, therefore, the effect on the shear velocity was considered negligible. Wall shear velocity obtained using the stated procedure was later compared with the empirical relationship from Smits et al. (1983) and the maximum difference was 2% for reference smooth wall measurements. Based on these considerations, it is possible to obtain the wall friction velocity within 2% accuracy. In order to improve the accuracy, sufficient data points in the viscous sub-layer is required.

## 2.3 Results

### 2.3.1 Results: Standard Boundary Layer validation

A reference measurement of TBL over smooth surface was measured prior to the measurements with perforated surface. Therefore, this data over smooth wall will be referred as Standard Boundary Layer, subsequent measurements over perforated surface will be presented with reference SBL data set. This reference data set also act as a the boundary layer characterization. Careful measurements of SBL profiles using 2D LDA provided us the u and v-component of the velocity vector and subsequently, statistics of both velocity components upto fourth order is presented here with their corresponding inner and outer scaling parameters. In addition, LES and HWA data from Eitel-Amor (2014) and Örlü and Schlatter (2013) respectively are also compared.

In Figure-2.12 (a), law of wall presentation of the mean streamwise velocity and wall distance is a common way of validating SBL data. Here, $\bar{u}^+ = \bar{u}/u_\tau$ is plotted against $y^+ = yu_\tau/\nu$ following Equation-1.18 and compared with the LES and HWA data for various Reynolds number as mentioned in the legend section. As the inset magnifies the viscous sub-layer e.g $y^+ \leq 5$, we can see an excellent agreement with the reference LES and HWA data compared. This includes the viscous sub-layer till the end of wake region where apparently, an excellent overlapping in the log region is observed.

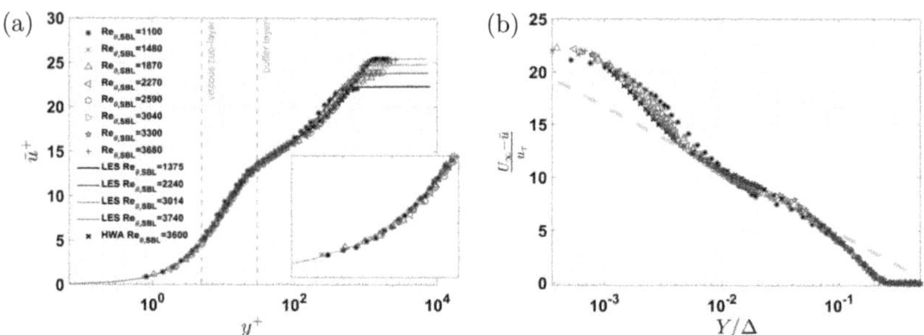


**Figure 2.12:** (a) Inner scaled law of wall presentation of mean streamwise velocity $(\bar{u}/u_\tau)$ against $(yu_\tau/\nu)$ for reference SBL data, every fourth data is plotted in the main figure in order to maintain clarity, vertical lines indicate the upper limit of viscous sub-layer and buffer layer respectively Wei et al. (2005). Inset figure magnifies the region $y^+$=1~5. (b) Outer scaled velocity defect profiles, *Dashed line* indicate *velocity defect-law* Equation-1.25

In order to check the outer length scaling, Figure-2.12 (b) shows the velocity defect law according to Coles (1956). The data is compared with log law expressed in the velocity-defect formulation e.g Equation-1.25, where $\eta = y/\Delta$. The constant B1=-1 which is slightly above the value, B1=-0.87 suggested by Monkewitz et al. (2007).

Alfredsson and Örlü (2010) and Alfredsson et al. (2011) proposed a new scaling of boundary layer profile data in order to avoid uncertainties that arises from the sensitive wall shear stress measurements. This scaling is based on the Equation-1.31 with the wake region ($\bar{u}/U_\infty \geq 0.71$) as can be seen in Figure-2.13(a). Alfredsson and Örlü (2010) proposed the values of constants in Equation-1.31 as a=0.031 and b=0.260, whereas fitting of present SBL data derives a=0.01 and b=0.240. However, outer region data has excellent overlapping and agreed well to Equation-1.31. $\bar{u}/U_\infty = 0.2$ refers to $y^+ \sim 5$ or viscous sub-layer and $\bar{u}/U_\infty = 0.15$ refers to $y^+ \sim 3$ which is fairly linear. Moreover, near wall data is not effected at larger Reynolds number for $\bar{u}/U_\infty \leq 0.15$. We can also see that max $u_{rms}$ is located at $\bar{u}/U_\infty = 0.42 \sim 0.44$ which corresponds to $y^+ = 10 \sim 11$.

Figure-2.13(b) is an alternative way to plot the same velocity data as in (a). This is plotted along the same idea of the diagnostic plot but here $\bar{u}/U_\infty$ is plotted as a function of $u_{rms}/\bar{u}$. The turbulence intensity increases linearly following Equation-1.32. For the region $\bar{u}/U_\infty \geq 0.6$ e.g outer region, the collapse of the data is excellent. We can also observe a linear plateau for buffer layer data although there is no collapse at all.

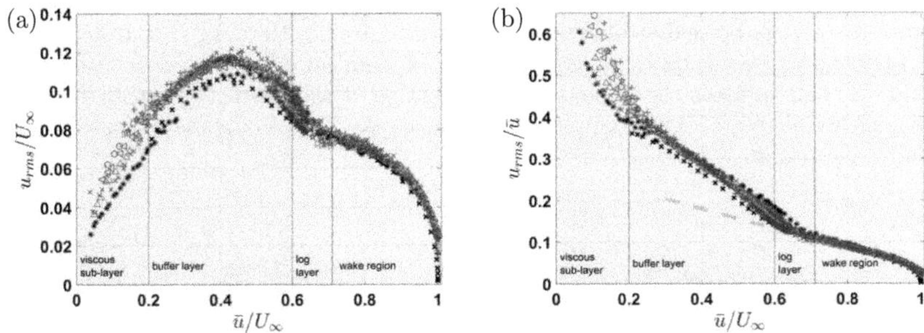


**Figure 2.13:** (a) *The diagnostic plot* (Alfredsson and Örlü (2010)) for $\bar{u}/U_\infty$ as a function of $u_{rms}/U_\infty$, here '- - - -' indicate Equation-1.31. (b) Streamwise turbulence intensity normalized with the local streamwise velocity as a function of $\bar{u}/U_\infty$, '- - - -' indicate Equation-1.32. Same data and symbols as in Figure-2.12(a).

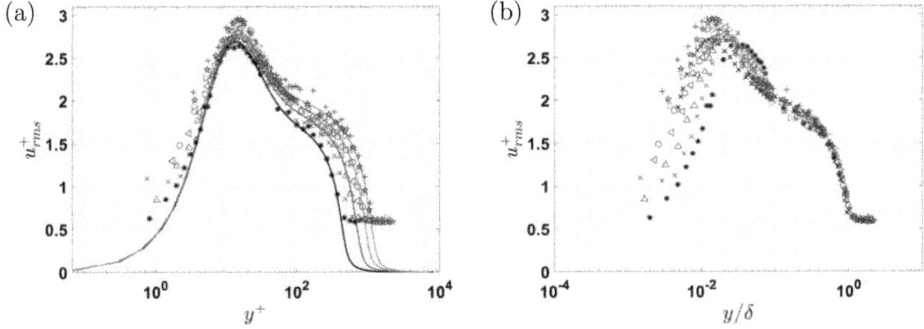


**Figure 2.14:** (a) Profiles of RMS values of streamwise velocity fluctuations as a function of wall distance using inner scaling. LDA data (markers) is compared with LES data (continuous line) from Eitel-Amor (2014) (b) Profiles of RMS values of streamwise velocity fluctuations using inner scaling as a function of $y/\delta$. Markers and continuous lines represent same data as in Figure-2.12(a). For clarity of the figure, one out of every 4 data is plotted.

The Root-Mean-Squared (RMS) streamwise fluctuation ($u^+_{rms} = u_{rms}/u_\tau$) profiles at all Reynolds number for SBL cases compared to the well resolved LES from Eitel-Amor (2014) are presented at Figure-2.14 (a). Location of so-called inner peak is located between $y^+$=10~11, although the inner peak is Reynolds number dependent. It is also observed that background turbulence intensity is slightly greater than 0.5%. However, Figure-2.14 (a) has presented both inner and outer peak with good spatial accuracy. Both LES and experimental data shows excellent agreement except the near wall data.

Figure-2.14 (b) present the same data as a function of wall distance normalized with outer length scale e.g $\delta$. This exhibit excellent collapse for the outer layer, specifically log and wake region.

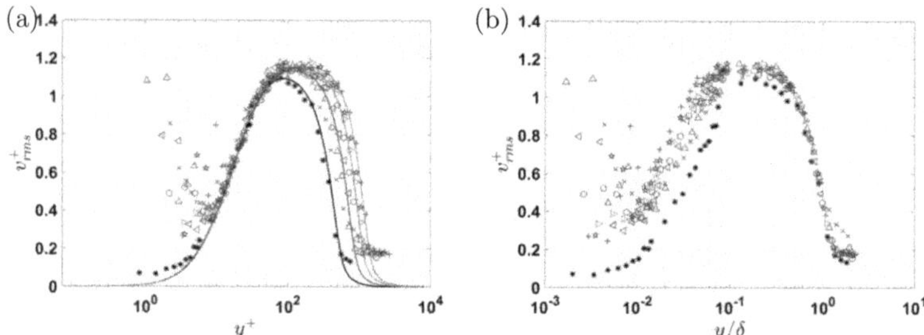


**Figure 2.15:** (a) Profiles of RMS values of wall-normal velocity fluctuations as a function of wall distance using inner scaling. LDA data (markers) is compared with LES data (continuous line) from Eitel-Amor (2014) (b) Profiles of RMS values of wall-normal velocity fluctuations using inner scaling as a function of $y/\delta$. Markers and continuous lines represent same data as in Figure-2.12(a). For clarity of the figure, one out of every four data is plotted.

Figure-2.15(a) shows $v^+_{rms} = v_{rms}/u_\tau$ as a function of $y^+$. Near wall data is effected by wall, atleast for $y^+ \leq 5$. However, excellent agreement to the LES data is observed starting from buffer region to the outer region. The same data also shows nice collapse for the outer region when plotted against $y/\delta$ (outer length scale) as shown in Figure-2.15(b).

Reference Hasanuzzaman et al. (2020) and Vallikivi et al. (2015a) showed from measurements in at high Reynolds number using PIV data, the skewness profiles of is well collapsed using inner coordinates over the region of $100 < y^+ < 0.15\delta^+$. They have also shown that skewness values are approaching negative values at $y^+ \approx 200$ before finally reach a value of $S \approx -0.1$. Finally, skewness values are more negative in the wake due to intermittency, where they collapse well with the outer coordinates.

We have also extended the statistics of the streamwise velocity data upto 3rd and 4th order moments e.g skewness ($S(u)$) and flatness ($F(u)$) following Equation-2.16 and 2.17 respectively. Figure-2.16(a) shows the skewness profiles of streamwise velocity plotted with inner and outer scales respectively for SBL cases. Although, Figure-2.16(a) exhibit Reynolds number dependence for the outer region when scaled with inner length scales. At all Reynolds number, skewness is slightly positive near the wall for $y^+ < 200$ and gradually becoming more negative as the distance from the wall increases. In order to compare the accuracy, LES and HWA results are also plotted. Although both figures exhibit Reynolds number dependence for the outer region when scaled with inner length scales, nevertheless, excellent agreement is observed with LES and HWA data for SBL data.

Figure-2.16(b), plots the flatness profiles with inner scaling. Inner coordinates shows Reynolds number dependence in the outer region but good convergence is visible at least for the viscous and logarithmic region. This figure is also in comparison to the numerical data has an excellent agreement.

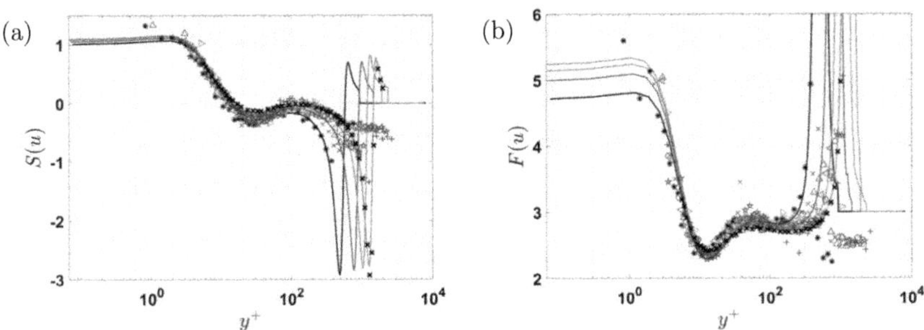


**Figure 2.16:** (a) Skewness and (b) Flatness of streamwise component normalized with inner scale parameters compared with LES and HWA data from Eitel-Amor (2014) and Örlü and Schlatter (2013) respectively as a function of inner scaled wall distance. Same data and symbols as in Figure-2.12(a). Markers used For clarity of the figure, one out of every four data is plotted.

Table-3.3 summarizes the mean and integral properties of the SBL cases presented in the section-2.3.1. Here, data is presented with rounding off error ≤1%.

### 2.3.2 Results: Integral properties

In Figure-2.17(a), friction coefficients ($C_f$) along $Re_{\theta,SBL}$ is plotted for different blowing ratios applied. Determining process of $C_f$ is previously explained in section-2.2.2. In order to make the friction reduction effect clearly visible, $C_f$ is plotted along the reference smooth wall $Re_{\theta,SBL}$ although $Re_\theta$ increases with BR. SBL data is compared with analytical relationship from Smits et al. (1983) and the experimental results of Zanoun et al. (2014) using 1D miniaturized LDA respectively. Results from Zanoun et al. (2014) is slightly underestimated then the analytical data from Smits et al. (1983). on the other hand, present data has less deviation from the analytical relationship. $C_f$ measurements from perforated surface without blowing exhibit slight variation from the SBL data. However, when blowing is applied at B1 velocity, $C_f$ is significantly reduced but as the blowing velocity remains unchanged compared to the increased $Re_{\theta,SBL}$, reduction of $C_f$ is unchanged compared to the analytical relationship. This is in general visible from $Re_{\theta,SBL}$=2270~3670.

$C_f$ rate of reduction in percentage along $Re_{\theta,SBL}$ is presented in Figure-2.17(b). It is observed that the rate of $C_f$ reduction is higher at low Reynolds number range, however, reduction decreases as the Reynolds number of the grows up. Starting from the $Re_{\theta,SBL}$=2270, this remains as a stable plateau but keeps growing at $Re_{\theta,SBL}$=3030 and until $Re_{\theta,SBL}$=3670. However, $C_f$ estimation is pretty sensitive at higher Reynolds number and suffers from larger uncertainty as the Reynolds number grows. This is due to the lack of data point in viscous sub-layer. This is visible in Figure-2.17 (c), where, wall correction ($y_0$) is plotted. Figure-2.17 (c) indicate the BR (2.1.2) relationship with increasing $Re_{\theta,SBL}$. BR gradually decreases with Reynolds number as the blowing velocities B1 and B2 is fixed. This effect of BR is particularly relevant in order to explain the reducing $C_f$ effect as Reynolds number grows in Figure-2.17 (a).

| $U_\infty$ (m/s) | $u_\tau$ (m/s) | $\nu/u_\tau$ ($\mu m$) | $C_f$ [×10¯3] | $\delta$ (mm) | $\delta^*$ (mm) | $\theta$ (mm) | H - | $Re_\theta$ | $Re_{\delta^*}$ | $Re_x$ [$\times10^6$] | $Re_\tau$ - |
|---|---|---|---|---|---|---|---|---|---|---|---|
| 10.12 | 0.461 | 32.61 | 4.143 | 13.543 | 2.4 | 1.64 | 1.44 | 1100 | 1593.2 | 0.381 | 416 |
| 13.55 | 0.594 | 25.35 | 3.8456 | 13.063 | 2.3 | 1.64 | 1.43 | 1480 | 2119.2 | 0.508 | 506 |
| 18.07 | 0.773 | 19.56 | 3.6599 | 12.112 | 2.2 | 1.56 | 1.42 | 1870 | 2658.9 | 0.675 | 605 |
| 23.23 | 0.973 | 15.59 | 3.5079 | 11.762 | 2.1 | 1.48 | 1.41 | 2270 | 3189.9 | 0.866 | 738 |
| 28.03 | 1.171 | 13.40 | 3.4919 | 11.304 | 2.0 | 1.44 | 1.39 | 2590 | 3604.6 | 1.01 | 811 |
| 32.68 | 1.317 | 11.64 | 3.248 | 10.989 | 2.0 | 1.42 | 1.39 | 3030 | 4214.9 | 1.2 | 912 |
| 36.43 | 1.444 | 10.65 | 3.1443 | 10.906 | 1.9 | 1.39 | 1.38 | 3300 | 4584.5 | 1.337 | 1024 |
| 4.55 | 1.758 | 8.90 | 3.1164 | 10.473 | 1.8 | 1.29 | 1.38 | 3670 | 5076.8 | 1.609 | 1160 |

**Table 2.3:** Mean properties (for reference cases) of the SBL in LAS wind tunnel at X = 58% obtained from LDV measurement. Here, $\delta^* = (\sum_{y=0}^{\infty}(1 - \bar{u}/U_\infty))$, displacement thickness; $\theta = \sum_{y=0}^{\infty}(\bar{u}/U_\infty)(1-\bar{u}/U_\infty)dy$, momentum thickness; $H = \delta^*/\theta$, shape factor; $Re_{\delta^*} = \delta^* U_\infty/\nu$, Reynolds number based on local displacement thickness; $Re_\tau = \delta u_\tau/\nu$, Reynolds number based on local friction velocity.

Figure-2.19(a) to (j) present the plots of integral properties at their corresponding

$Re_{\theta,SBL}$. Figure-2.19(a), (b) and (c) exhibit dimensional $\delta$, $\delta^*$ and $\theta$ respectively (which indicate the outer length scale, displacement length and momentum loss length respectively). These integral parameters are increased due to blowing. In general,these properties are enhanced as the blowing increases. It is observed that these length scales are strongly influenced by the rate of blowing. At the lowest boundary of present Reynolds number range, effect is more prominent. Therefore, effect is gradually attenuated as the Reynolds number increases. This can be explained by recalling the Figure-2.17(d), where it was shown that how BR varies at fixed $V_W$=B1,B2 and B3 for varying Reynolds number.

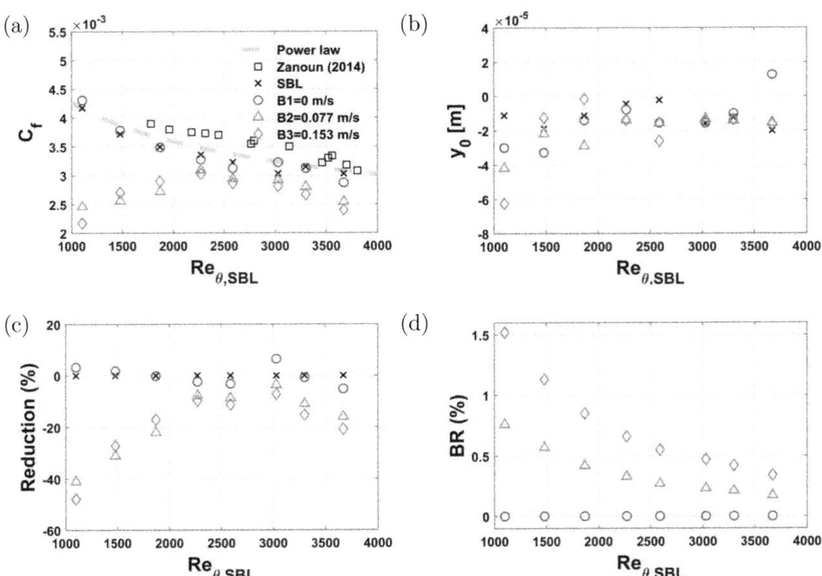


**Figure 2.17:** (a) $C_f$ against reference $Re_{\theta,SBL}$, Power law data is indicated from the relation $C_f Re_{\theta,SBL} = K$ (Smits et al. (1983)) for smooth wall data where constant, K=0.024. (b) Wall correction ($y_0$) in dimensional form, (c) Reduction of $C_f$ in percentage and (d) Blowing ratio (BR) variation with the $Re_{\theta,SBL}$.

Similarly, Figure-2.19 (e) and (f) present the shape factor ($H$) and rate of change with blowing ratio for each case investigated respectively. In addition, this is also compared with the data from Örlü and Schlatter (2013). The shape factor increases with blowing ratios for all cases. As for the momentum thickness, the shape factor increases with blowing rate with same ratio compared to reference case for all Reynolds number investigated (see Figure-2.19(f). Combining the observations on $\theta$ and $H$, the increase of blowing ratio determined the increase of boundary layer thickness ($\delta$) independently of the Reynolds number studied here. The difference between rate of change of shape factor for SBL and perforated surface without blowing is negligible as inside the uncertainty level. Therefore, it can be inferred that effect of roughness from perforated surface is also negligible at least for the range of Reynolds number measured in the present paper. Kametani and Fukagata (2011) and Kametani et al. (2015) also suggested that blowing increases the boundary layer thickness, momentum thickness and the shape factor. Although, present measurement agrees to the DNS result trend of the mentioned papers, DNS overestimate the growth rate of the stated mean properties. There are two possibilities of such deviation: firstly, growth rate of mean properties reduces with increased blowing or secondly, blowing effect is more prominent at low Reynolds number. The second hypothesis can be privileged as the growth rate of $\theta$ and $H$ with BR is nearly linear and independent of momentum Reynolds number Hasanuzzaman et al. (2020).

Figure-2.19 (g) plots the Reynolds number based on momentum thickness to the reference Reynolds number obtained from SBL cases. As it was observed from Figure-2.19 (c) where the growth of momentum thickness is plotted is also similarly reflected here. Consequently, the rate of growth for $Re_\theta$ is also presented in Figure-2.19 (h).

Figure-2.19 (i) presents the Reynolds number based on displacement thickness along $Re_{\theta,SBL}$. This follows the trend observed in Figure-2.19 (b).

Although it was presented in Figure-2.17 (c), the rate of friction co-efficient reduction, the similar trend is also observed in Figure-2.19 (j), where $Re_\tau$ is plotted along reference Reynolds number.

### 2.3.3 Results: Statistics

Figure-2.20(a) to (h), present the mean streamwise velocity profiles for different Reynolds number scaled with $U_\infty$ where as wall distance is normalized with $\delta$. Here, inset figures magnifies the region $\bar{u}/U_\infty = 0.3 \sim 0.9$ and $y/\delta = 0.005 \sim 0.55$. Data points are plotted as lines in order to remove ambiguity and for better clarity of the profiles. We can observe, a significant modification of the profiles depending on the BR. Therefore, mean streamwise velocity is suppressed and the suppression increases as BR increases. When this is compared for different Reynolds numbers, this effect is quite persistent although BR is comparatively small in magnitude at higher Reynolds number. The location of the modification is fairly at the same location for all Reynolds number investigated.

Figure-2.21(a) to (h), is the same plot scaled with viscous length scales in order to see the law of wall presentation of the profiles with different BR. In addition, SBL profiles along with log (Equation-1.18) and linear (Equation-1.28) respectively. Most recently, inner scaled profiles of streamwise velocity under the influence of uniform blowing was reported by Kametani et al. (2015). They have described that the TBL flow under

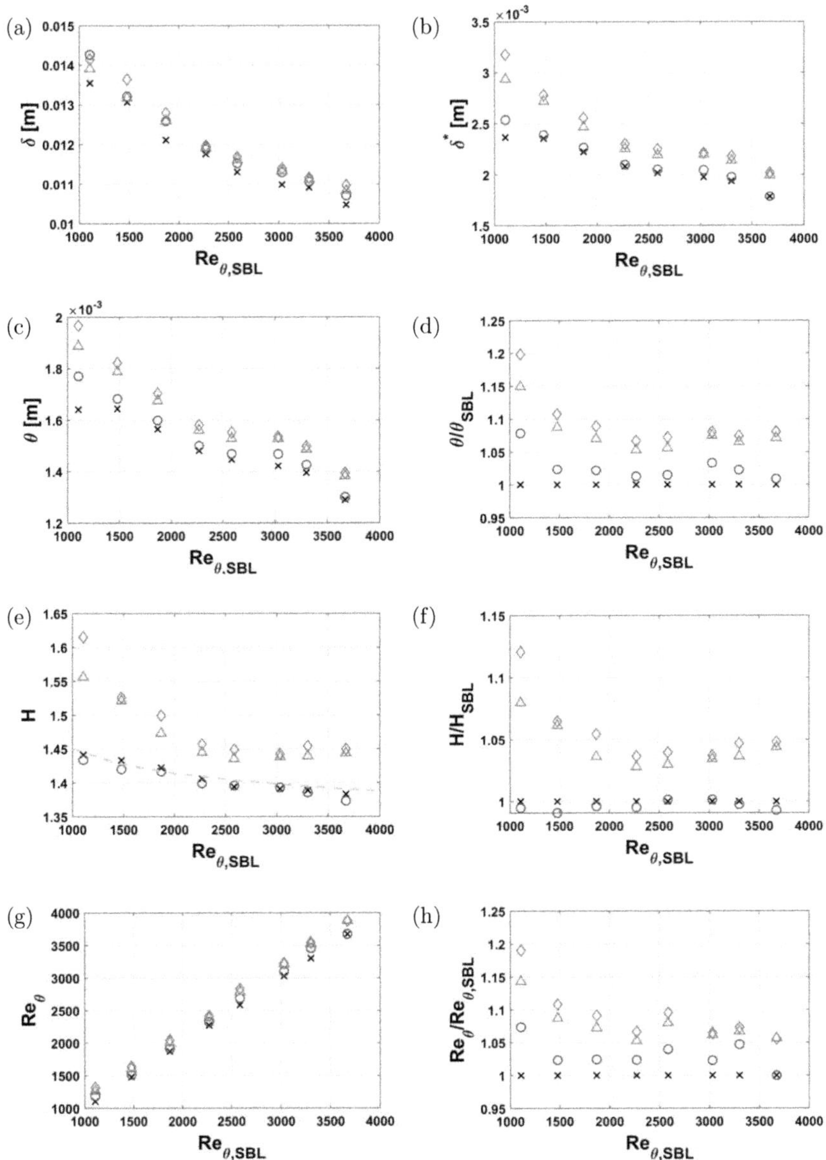
(a)
δ [m]
Re$_{θ,SBL}$
(b)
δ* [m]
Re$_{θ,SBL}$
(c)
θ [m]
Re$_{θ,SBL}$
(d)
θ/θ$_{SBL}$
Re$_{θ,SBL}$
(e)
H
Re$_{θ,SBL}$
(f)
H/H$_{SBL}$
Re$_{θ,SBL}$
(g)
Re$_θ$
Re$_{θ,SBL}$
(h)
Re$_θ$/Re$_{θ,SBL}$
Re$_{θ,SBL}$

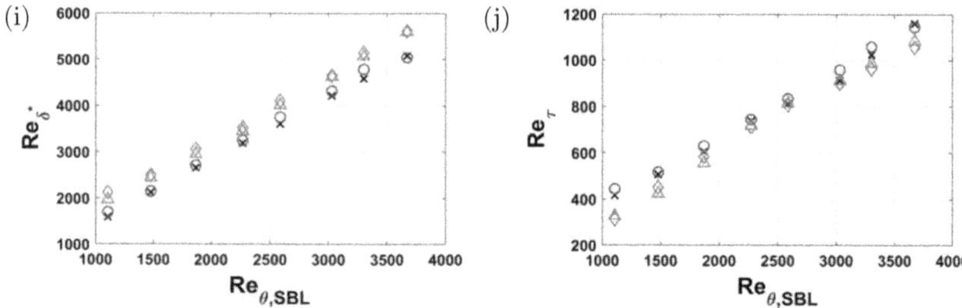


**Figure 2.19:** plots of integral properties as a function of $Re_{\theta,SBL}$ where markers at different blowing ratios are used similar to the Figure-2.17; (a) Boundary layer thickness, (b) Displacement thickness ,(c) Momentum thickness, (d) change of momentum thickness in comparison to SBL cases, (e) Shape factor, '– –' indicate Örlü and Schlatter (2013), (f) change of shape factor in comparison to SBL cases, (g) Momentum thickness Reynolds number, (h) change of Momentum thickness Reynolds number, (i) Displacement thickness Reynolds number, (j) friction Reynolds number.

uniform blowing[10] will deviate from the log law (Equation-3.6) when corresponding inner scaled parameters are used. For different BR applied, apparently, there is no changes observed for the streamwise velocity profiles in the viscous sub-layer ($y^+ \leq 5$). Even at higher BR at the lowest Reynolds number case (see Figure-2.21(a)), viscous layer exhibit almost no changes. However, mean profiles show strong deviation starting from the buffer till the wake region as can be seen in Figure-2.21(a) and (b).

The magnitude of the deviation depends strongly on the BR as we know that BR is larger for the lower boundary of the Reynolds numbers investigated, so as the mean profile deviation. As the Reynolds number keep increasing in Figure- 2.21(c), (d), (e) and (f), effect of blowing gradually diminishes in the near-wall region and shifted gradually to the outer region and therefore, the profiles have stronger modification in the wake region. Finally, MBT effects mostly the wake region at the higher boundary of the Reynolds number investigated. Figure-2.21 (g) and (h) exhibit the highest Reynolds number investigated. Although, BR is fairly low for the upper boundary of the Reynolds number, wake region responses strongly. Therefore, we see a shift of the modification at the outer edges of the mean profiles as the Reynolds number grows.

We have seen a strong deviation from the logarithmic law while plotting the mean streamwise velocity in semi-logarithmic axis. Therefore, Figure-2.22 (a) to (h) presents

[10] In order to realize uniform blowing surface as implemented in the numerical studies, requires finite surfaces with holes drilled at uniform distance. Therefore, one has to implement such surfaces in the experimental studies for an wide range of Reynolds number e.g MBT surfaces used in the present experiment

the *velocity defect-law* presentation of the mean streamwise velocity profiles along outer scaled wall distance ($y/\delta$). Deviation from the reference data is stronger with the increased BR. However, outer region overlaps in the near wake region. The merging of the profiles is commonly observed at $y/\delta \approx 0.5$ and this is observed at all Reynolds number investigated. It is needless to say that the behaviour is in agreement as observed in Figure-2.20.

Figure-2.23(a) to (h) presents the *diagnostic plot* as proposed by Alfredsson and Örlü (2010) where $\bar{u}/U_\infty$ is plotted as a function of $u_{rms}/U_\infty$. Although, strong deviation in the viscous sub-layer is visible, wake region shows good overlapping with the reference profiles irrespective of the BRs.

However, general observation regarding the friction effect of the perforated surface as can be seen from Figure-2.21 (a) to (h), the perforated surface used for blowing shows negligible or no deviation when compared to the SBL cases. Friction effect is negligible at the lowest Reynolds numbers but the effect is gradually increasing as the Reynolds number grows. This is in agreement of the observations from Hasanuzzaman et al. (2020).

Figure-2.24 (a) to (h) plots the profiles of inner scaled Root-Mean-Square for streamwise velocity fluctuation. Figure-2.24 (b), (d), (f) and (h) is compared with the nearest LES results and a good agreement is observed. SBL and perforated surface without blowing shows fairly similar profiles for inner region, nevertheless, outer peak is slightly enhanced. This deviation for the outer peak increases as the Reynolds number of the flow grows. For Figure-2.24 (a) and (b), profiles with different blowing velocity do not collapses each other, this indicates that the BR is comparatively higher compared to the Reynolds number and therefore, boundary layer profiles are strongly influenced upto the edge of the boundary. Hence, profiles are not recovered in terms of their RMS fluctuation. But as the Reynolds number grows e.g starting at the $Re_{\theta,SBL}$ = 1870, profiles recovers themselves in terms of the blowing velocity. However, we can observe for all the Reynolds number investigated, an inherited background turbulence intensity of the wind tunnel in the order of 0.5%.

At all Reynolds number investigated, inner peak location is found $y^+ = 10 \sim 11$. This inner peak location is valid for different blowing velocity B1, B2 and B3 as well. However, although inner peak is qualitatively similar, outer layer peak or the outer peak is strongly influenced due to blowing. This is even enhanced with increased blowing velocity which can be seen at all Reynolds number.

Figure-2.25 (a) to (h) compares the inner scaled skewness profiles of streamwise velocity for different Reynolds number investigated. In Figure-2.25 (b), (d), (f) and (h), skewness profiles are compared to the LES results and found in excellent agreement. In general, skewness profiles with different blowing velocity is qualitatively identical. We can observe a positive skewness for the region $y^+ \leq 10$. However, wall region $10 \leq y^+ \leq 200$ shows the strongest deviation of the skewness. For all the Reynolds number investigated, it varies from 1/2 to -1/2. Skewness tends to move into positive direction as the blowing velocity increases. For $y^+ \geq 200$, skewness profiles with different blowing velocity and SBL cases are comparable.

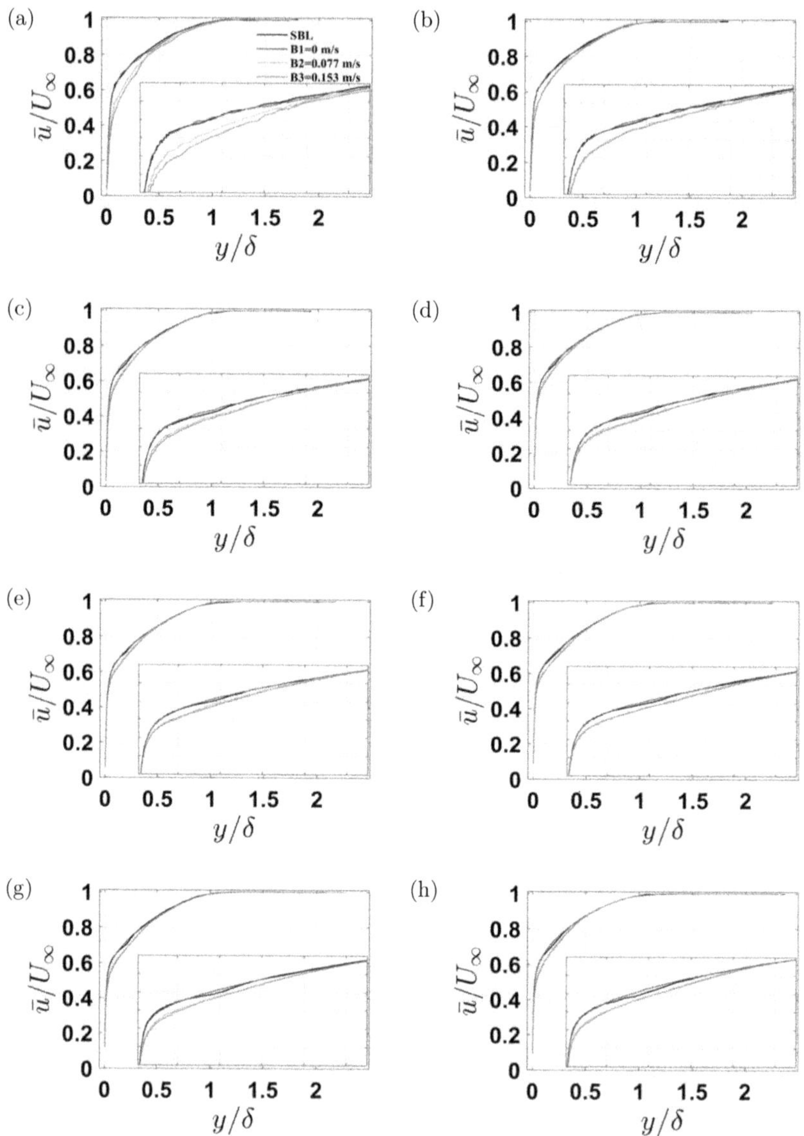


**Figure 2.20:** plots of mean streamwise velocity profiles using outer length scales. Inset figures magnifies the region $\bar{u}/U_\infty = 0.3 \sim 0.9$ and $y/\delta = 0.005 \sim 0.55$. (a) $Re_{\theta,SBL} = 1100$, (b) $Re_{\theta,SBL} = 1480$, (c) $Re_{\theta,SBL} = 1870$, (d) $Re_{\theta,SBL} = 2270$, (e) $Re_{\theta,SBL} = 2590$, (f) $Re_{\theta,SBL} = 3030$, (g) $Re_{\theta,SBL} = 3300$, (h) $Re_{\theta,SBL} = 3670$.

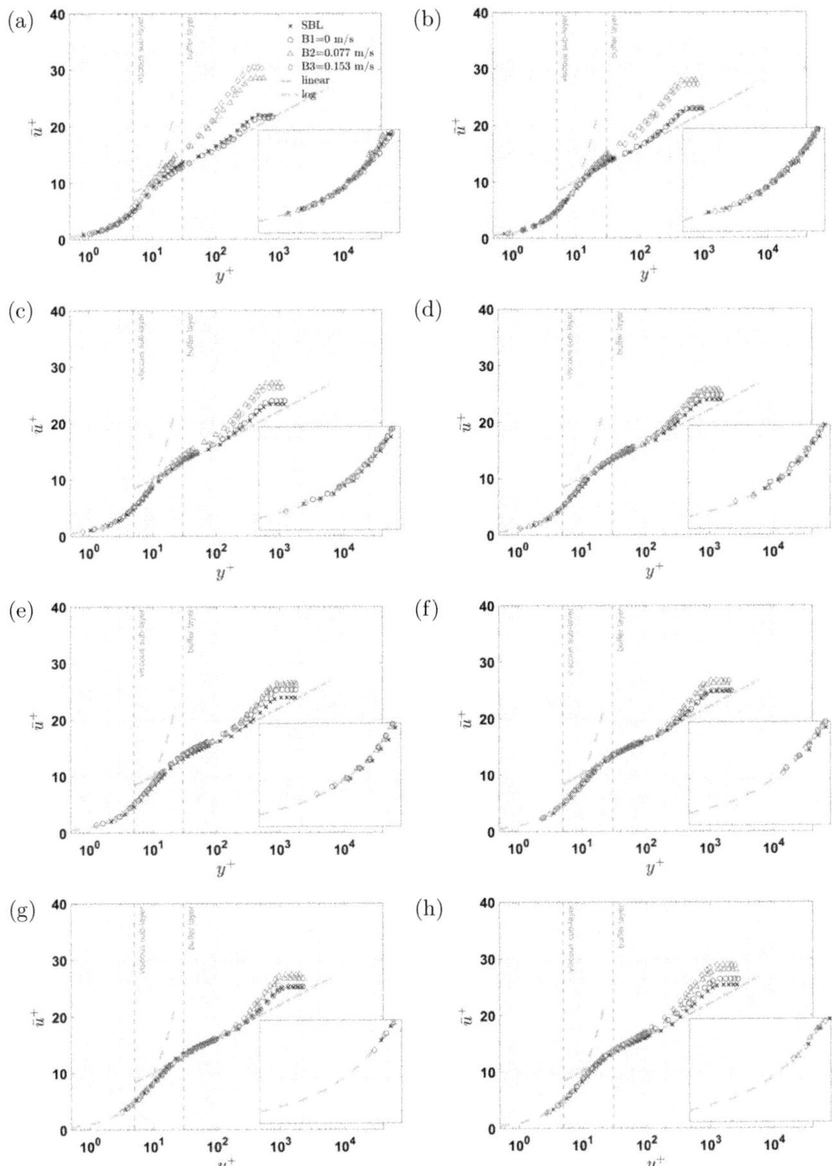


**Figure 2.21:** Inner scaled profiles of mean streamwise velocity, one out of every four data points are plotted, inset plot magnifies the region $y^+$ = 0.5 ~ 5 and $\bar{u}^+$ = 0 ~ 4.5, legend is similar for all plots as indicated in (a); (a) $Re_{\theta,SBL}$ = 1100, (b) $Re_{\theta,SBL}$ = 1480, (c) $Re_{\theta,SBL}$ = 1870, (d) $Re_{\theta,SBL}$ = 2270, (e) $Re_{\theta,SBL}$ = 2590, (f) $Re_{\theta,SBL}$ = 3030, (g) $Re_{\theta,SBL}$ = 3300, (h) $Re_{\theta,SBL}$ = 3670.

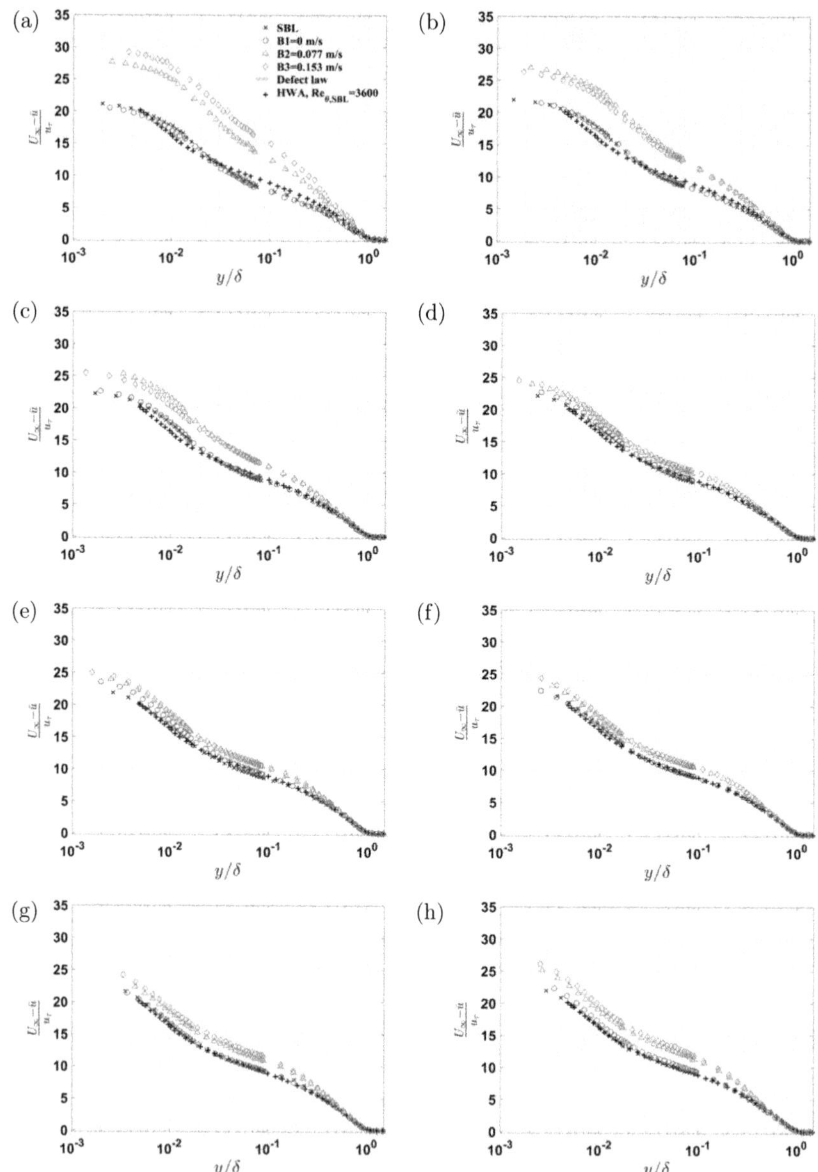


**Figure 2.22:** *Velocity defect-law* presentation of mean streamwise velocity, one out of every four data points are plotted, legend is similar for all plots as indicated in (a); (a) $Re_{\theta,SBL}$ = 1100, (b) $Re_{\theta,SBL}$ = 1480, (c) $Re_{\theta,SBL}$ = 1870, (d) $Re_{\theta,SBL}$ = 2270, (e) $Re_{\theta,SBL}$ = 2590, (f) $Re_{\theta,SBL}$ = 3030, (g) $Re_{\theta,SBL}$ = 3300, (h) $Re_{\theta,SBL}$ = 3670.

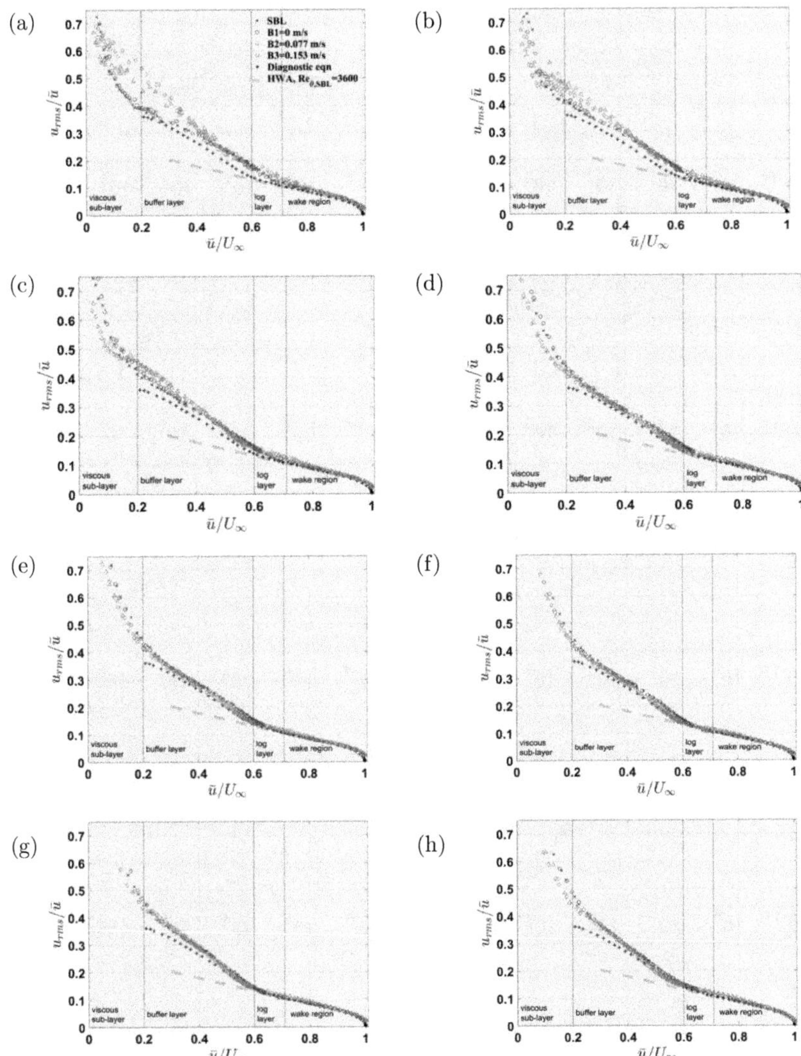


**Figure 2.23:** *Diagnostic plot* presentation of mean streamwise velocity, '- - - -' indicate Equation-1.31. One out of every four data points are plotted, legend is similar for all plots as indicated in (a); (a) $Re_{\theta,SBL}$ = 1100, (b) $Re_{\theta,SBL}$ = 1480, (c) $Re_{\theta,SBL}$ = 1870, (d) $Re_{\theta,SBL}$ = 2270, (e) $Re_{\theta,SBL}$ = 2590, (f) $Re_{\theta,SBL}$ = 3030, (g) $Re_{\theta,SBL}$ = 3300, (h) $Re_{\theta,SBL}$ = 3670.

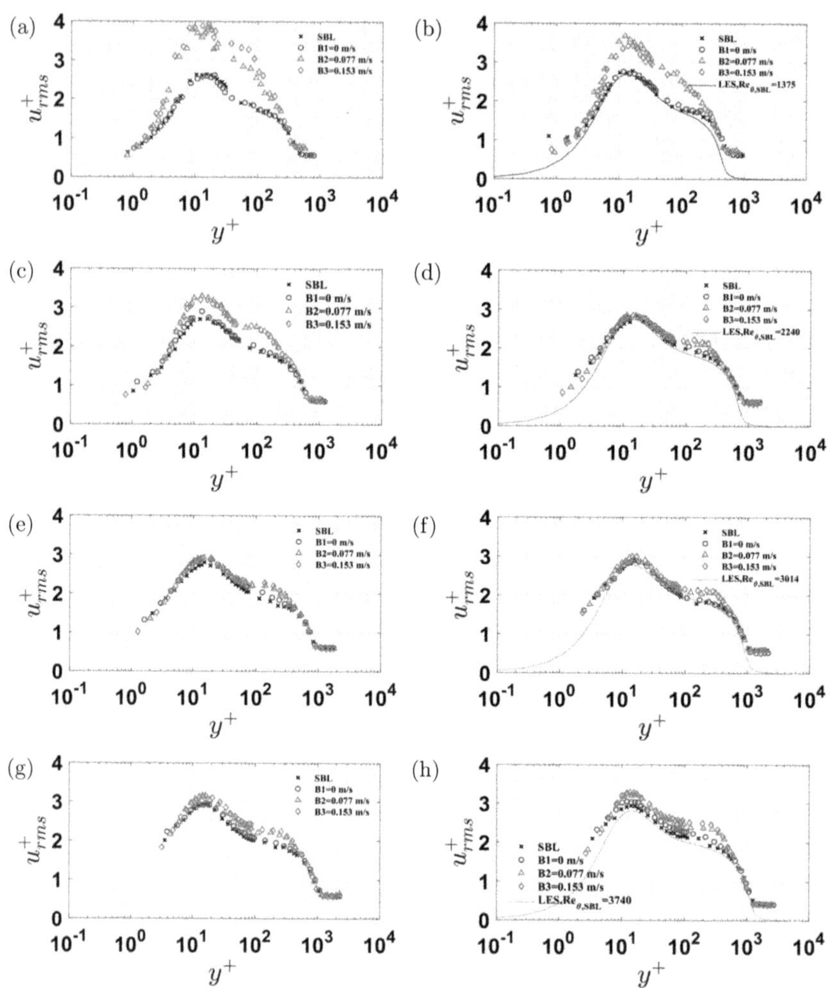


**Figure 2.24:** Inner scaled RMS of mean streamwise velocity profiles for individual Reynolds number, one out of every four data points are plotted; for (a) $Re_{\theta,SBL}$ = 1100, (b) $Re_{\theta,SBL}$ = 1480, (c) $Re_{\theta,SBL}$ = 1870, (d) $Re_{\theta,SBL}$ = 2270, (e) $Re_{\theta,SBL}$ = 2590, (f) $Re_{\theta,SBL}$ = 3030, (g) $Re_{\theta,SBL}$ = 3300, (h) $Re_{\theta,SBL}$ = 3670.

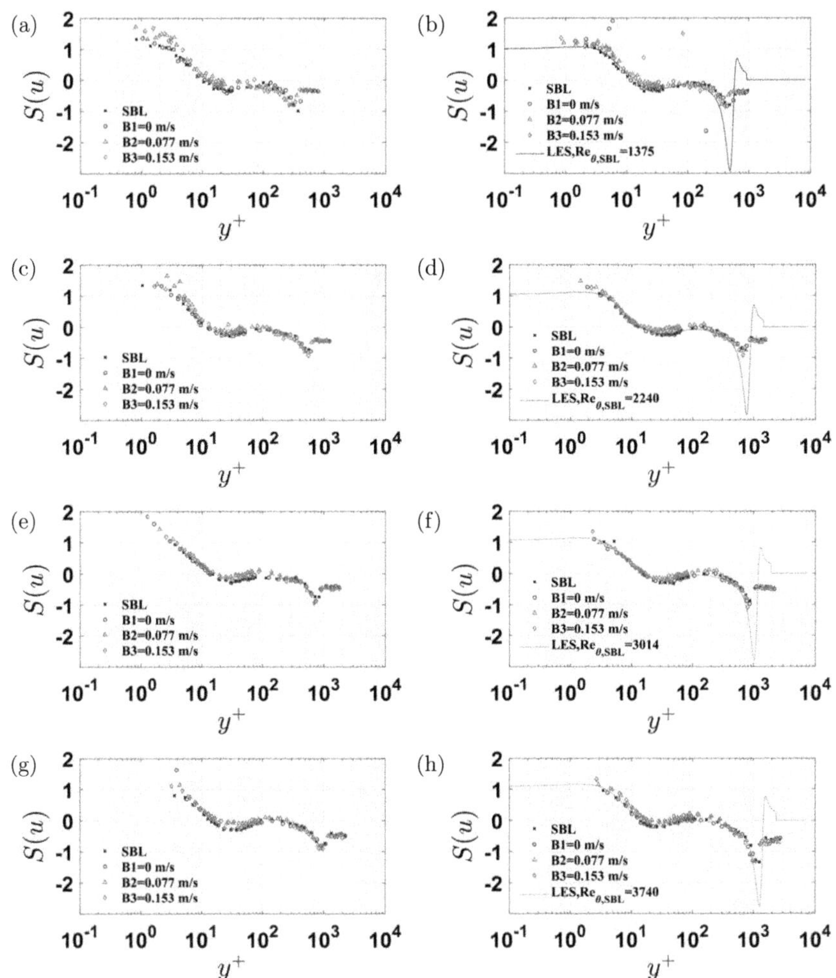


**Figure 2.25:** Skewness profiles of mean streamwise velocity for individual Reynolds number, one out of every four data points are plotted; for (a) $Re_{\theta,SBL}$ = 1100, (b) $Re_{\theta,SBL}$ = 1480, (c) $Re_{\theta,SBL}$ = 1870, (d) $Re_{\theta,SBL}$ = 2270, (e) $Re_{\theta,SBL}$ = 2590, (f) $Re_{\theta,SBL}$ = 3030, (g) $Re_{\theta,SBL}$ = 3300, (h) $Re_{\theta,SBL}$ = 3670.

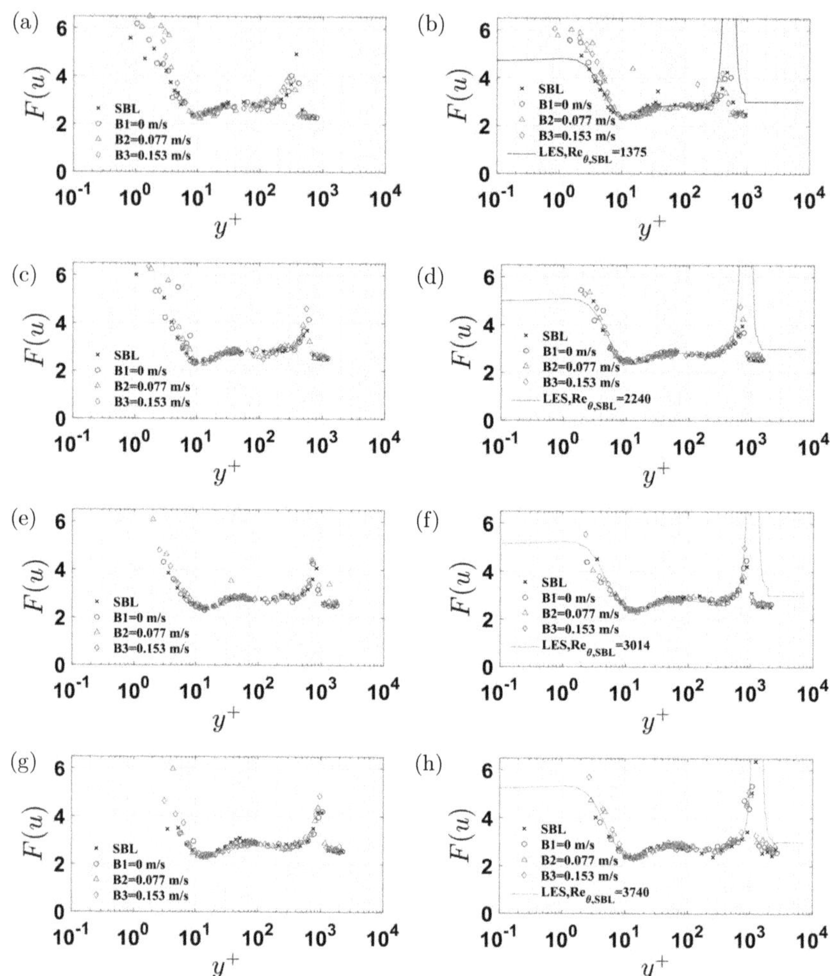


**Figure 2.26:** Flatness profiles of mean streamwise velocity for individual Reynolds number, one out of every four data points are plotted; for (a) $Re_{\theta,SBL}$ = 1100, (b) $Re_{\theta,SBL}$ = 1480, (c) $Re_{\theta,SBL}$ = 1870, (d) $Re_{\theta,SBL}$ = 2270, (e) $Re_{\theta,SBL}$ = 2590, (f) $Re_{\theta,SBL}$ = 3030, (g) $Re_{\theta,SBL}$ = 3300, (h) $Re_{\theta,SBL}$ = 3670.

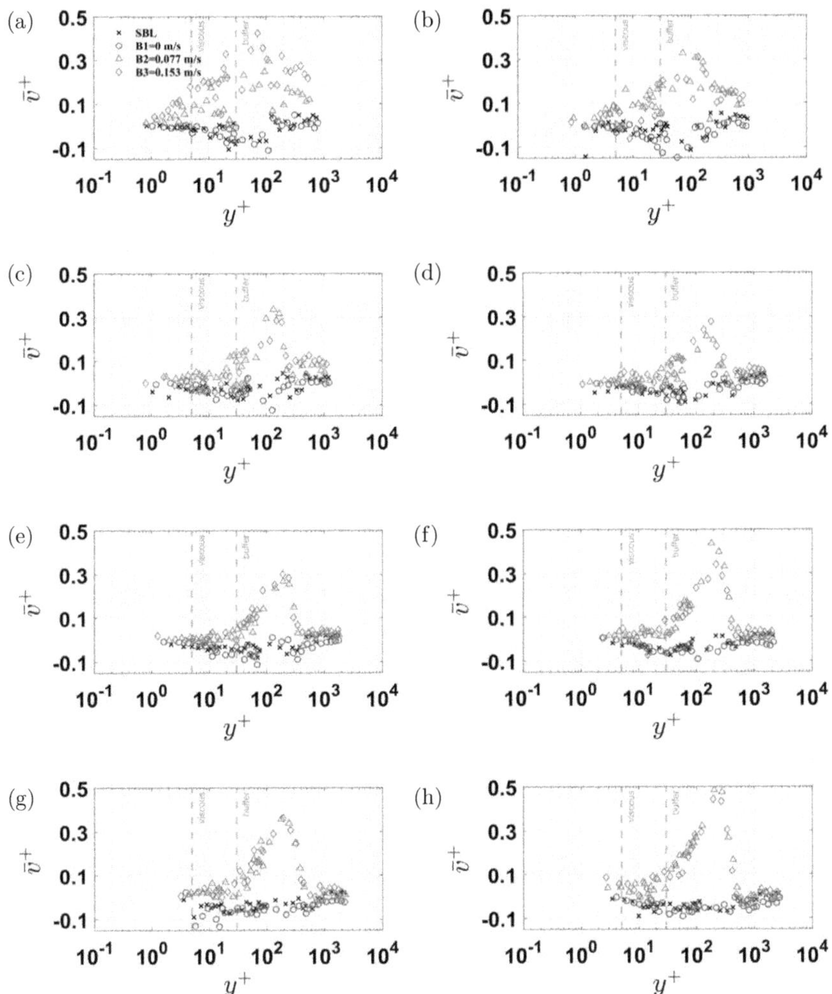


**Figure 2.27:** Inner scaled profiles of mean wall-normal velocity for individual Reynolds number, one out of every four data points are plotted; for (a) $Re_{\theta,SBL}$ = 1100, (b) $Re_{\theta,SBL}$ = 1480, (c) $Re_{\theta,SBL}$ = 1870, (d) $Re_{\theta,SBL}$ = 2270, (e) $Re_{\theta,SBL}$ = 2590, (f) $Re_{\theta,SBL}$ = 3030, (g) $Re_{\theta,SBL}$ = 3300, (h) $Re_{\theta,SBL}$ = 3670.

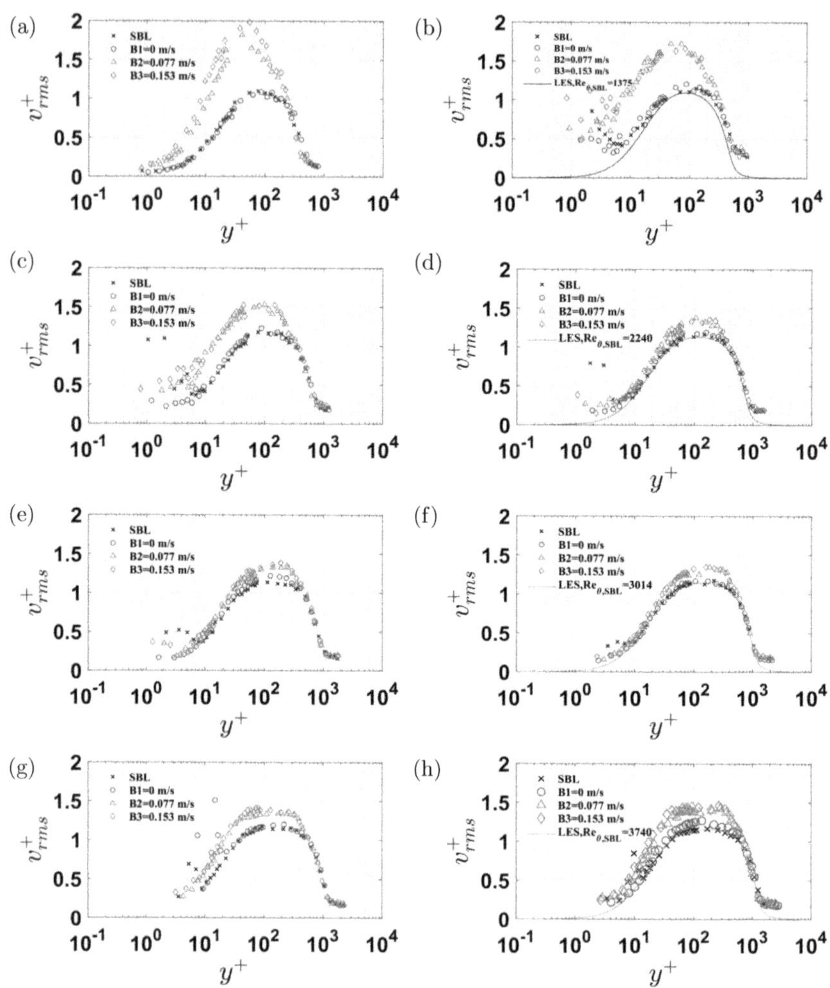


**Figure 2.28:** RMS profiles of mean wall-normal velocity for individual Reynolds number, one out of every four data points are plotted; for (a) $Re_{\theta,SBL}$ = 1100, (b) $Re_{\theta,SBL}$ = 1480, (c) $Re_{\theta,SBL}$ = 1870, (d) $Re_{\theta,SBL}$ = 2270, (e) $Re_{\theta,SBL}$ = 2590, (f) $Re_{\theta,SBL}$ = 3030, (g) $Re_{\theta,SBL}$ = 3300, (h) $Re_{\theta,SBL}$ = 3670.

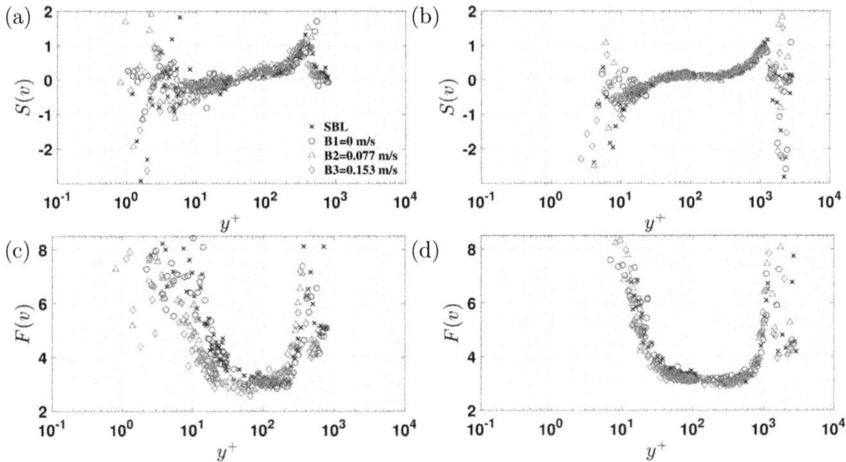


**Figure 2.29:** Skewness profiles of wall-normal velocity for (a) at $Re_{\theta,SBL}$ = 1100; (b) at $Re_{\theta,SBL}$ = 3670 and flatness profiles for (c) at $Re_{\theta,SBL}$ = 1100; (d) at $Re_{\theta,SBL}$ = 3670. One out of every four data points are plotted. Legend entry from (a) is valid for (b), (c) and (d)

Figure-2.26 (a) to (h) presents the flatness of the same data. When compared with their corresponding inner scaling, the profiles are similar in terms of different Reynolds number as well as for different blowing velocities. In other words, flatness is qualitatively uninfluenced due to different blowing velocities and this similarity is also observed at different Reynolds number.

Figure-2.27 (a) to (h) present the inner scaled mean wall normal velocity profiles. Gray dotted lines indicate limits of viscous sub-layer and buffer region respectively. SBL and B1 cases exhibits similar profiles, B2 and B3 are however, shows enhanced magnitude. Figure-2.27 (a) to (b) presents the lowest two Reynolds numbers where mean profile is also enhanced in the near wall region. A clear peak is visible at all Reynolds numbers. As the Reynolds number grows up for (c) to (h), blowing effect is more pronounced in the log region and reaches a peak value at $y^+ \approx 100 \sim 200$. Wall distance of the peak gradually slides away from the wall as the Reynolds number increases.

Figure-2.29 (a) and (b) plots the skewwness profiles of the mean wall normal velocity at the lowest and the highest Reynolds number respectively. Similarly, Figure-2.29 (c) and (d) plots the flatness. However, qualitatively no significant changes are visible in both the quantities.

Figure-2.28 (a) to (h) plots the inner scaled RMS of the wall normal velocity component. In Figure-2.28 (b), (d), (f) and (h), profiles are compared to the LES results and found in a good agreement. When SBL profiles are compared to the profiles over perforated surface without blowing e.g at B1, negligible deviation is found. Wall region $y^+ \leq 600$ shows an enhanced profile which gradually merges with the profiles similar to SBL case. Therefore, profiles overlap within $400 \leq y^+ \leq 600$ depending on the Reynolds number of the flow.

### 2.3.4 Results: performance indicator

When the final goal here is to apply FCT in order to reduce friction drag, therefore, a performance indicator analysis is required. In Figure-2.17(a) shows significant reduction in $C_f$, but this comes at a cost of energy in order to force air through the porous surface which is expensive. Within this context, it would be more rational to show a true drag reduction based on a control volume analysis and taking into account the work needed that eventually leads to drag reduction.

$$R = \frac{P_0 - P}{P_0} = \frac{C_{f,s} - C_{f,b}}{C_{f,s}} \tag{2.21}$$

$$S = \frac{(P_0 - (P + P_{in}))}{P} = \frac{C_{f,s} - (C_{f,b} + W_{in}/L_b)}{C_{f,b}} \tag{2.22}$$

$$W_{in} = \int \int_0^{Lb} [(P_w - P_{-w})V_W + 0.5V_W^3] \partial x \partial z \tag{2.23}$$

$$G = \frac{P_0 - P}{P_{in}} = \frac{C_{f,s} - C_{f,b}}{W_{in}/L_b} \tag{2.24}$$

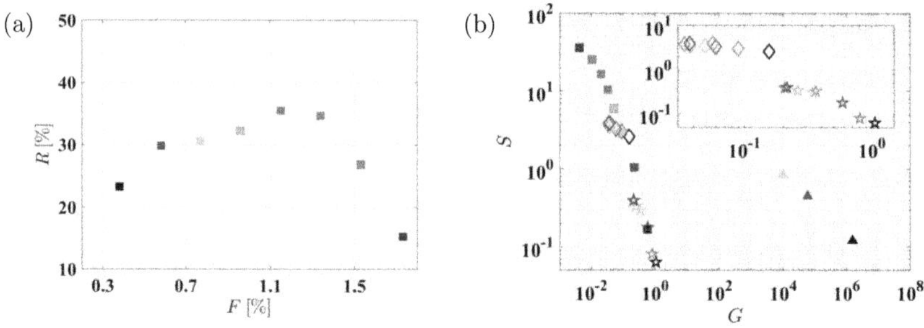


**Figure 2.30:** (a) Skin friction drag reduction rate (R) compared to the blowing ratio, (b) Net energy saving rate obtained by different BR in S-G map. Inset figure in (b) magnifies the data obtained during present experiment.

| symbols | source | BR [%] | Reynolds number | Exp./Num. |
|---|---|---|---|---|
| '■' | Motuz (2014) | 0 | $Re_{x,SBL} = 0.8 \times 10^6$ | Exp. (LDA) |
| '■' | - | 0.19 | - | - |
| '■' | - | 0.38 | - | - |
| '■' | - | 0.77 | - | - |
| '■' | - | 0.77 | - | - |
| '■' | - | 1.15 | - | - |
| '■' | - | 1.53 | - | - |
| '■' | - | 1.73 | - | - |
| '▲' | Kametani et al. (2015) | 0.1 | $Re_{\theta,SBL} = 2500$ | LES |
| '▲' | - | 0.5 | - | - |
| '▲' | - | 1 | - | - |
| '☆' | Present | 0.76 | $Re_{\theta,SBL} = 1100$ | Exp. (LDA) |
| '◊' | - | 1.52 | - | - |
| '☆' | - | 0.57 | $Re_{\theta,SBL} = 1480$ | - |
| '◊' | - | 1.13 | - | - |
| '☆' | - | 0.42 | $Re_{\theta,SBL} = 1870$ | - |
| '◊' | - | 0.85 | - | - |
| '☆' | - | 0.33 | $Re_{\theta,SBL} = 2270$ | - |
| '◊' | - | 0.66 | - | - |
| '☆' | - | 0.27 | $Re_{\theta,SBL} = 2590$ | - |
| '◊' | - | 0.55 | - | - |
| '☆' | - | 0.23 | $Re_{\theta,SBL} = 3030$ | - |
| '◊' | - | 0.47 | - | - |
| '☆' | - | 0.21 | $Re_{\theta,SBL} = 3300$ | - |
| '◊' | - | 0.42 | - | - |
| '☆' | - | 0.17 | $Re_{\theta,SBL} = 3670$ | - |
| '◊' | - | 0.34 | - | - |

**Table 2.4:** Symbol meanings in Figure-2.30 (a) and (b).

Efficiency of a certain FCT can be described in several ways ranging from the global to local input and output variables. Fukagata et al. (2002), Fukagata et al. (2009) and Kasagi et al. (2009b), proposed control performance indices in the wall bounded turbulent flow with active control (in Pipe, channel and control algorithm respectively). Particularly, Kasagi et al. (2009b) proposed most effective Control Performance Indicator's (CPI's) for wall bounded flows where external energy input is required, namely Drag Reduction Rate (R), Net Energy Saving Rate (S) and Gain (G). These indices were better explained for DNS results from ZPGTBL by Kametani and Fukagata (2011) at $Re_{\mathcal{T}} = 160$. The mentioned performance indicators are function of local friction coefficient ($C_f$), wall normal magnitude of blowing velocity ($V_W$) and the total work input due to air pumping ($W_{in}$). In our experiments, local friction coefficient was measured both on smooth and perforated surface at the same streamwise location, whereas blowing ratio was varied between 0.37% ≤ BR ≤ 1.73% for the measurements over perforated surface.

In Equation-2.21, R is equivalent to the reduction of pumping power and subscript 0 represent the cases without blowing (smooth surface data). Here, $C_{f,s}$ and $C_{f,b}$ correspond to the dimensionless friction co-efficient from smooth surface and surface with blowing.

For turbulent flow over flat plate at shear Reynolds number $Re_{\tau,SBL}$ = 812, varying the dimensionless blowing ratio from 0.37% to 1.73% to the corresponding drag reduction rate (R [%]) is presented in Figure-2.30 (a). On the other hand, by taking into account the power consumption ($P_{in}$), net energy saving rate (S) by the scheme is defined as Equation-2.22.

Here, $W_{in}$ is the input work directly obtained from the blowing velocity neglecting all the losses from pump through the transmission lines and $L_b$ is the stream-wise length of blowing surface and considered only the unit length of the blowing surface. Hence, the input power for blowing can be derived using Equation-2.23. Here, the first term on the right hand side of Equation-2.23 indicates the mean pressure on both sides and $V_W$ indicates the wall normal velocity from surface, this parameter is obtained from the control volume analysis of the flow rate ($Q_b$). From experimental validation we can neglect the pressure difference on both sides. Hence, input work per unit area is only a function of the perforated surface geometry and blowing velocity. Gain (G) in the overall system performance is expressed in Equation-2.23 and in wall bounded flows as in Equation-2.24 respectively. As discussed in sub-section-2.2.3, uncertainty regarding wall shear velocity is maximum 2% and hence, influence of this uncertainty for the estimation of friction co-efficient is 3.96%. The uncertainty from friction coefficient estimation is fairly constant for all measurements, therefore, the equations used to derive performance indicators (R, S and G) can successfully avoid this uncertainty.

Figure2.30(a), presents the drag reduction rate (R, Equation-2.21) to their corresponding blowing ratio at $Re_{\tau,SBL}$ = 812. For increasing blowing ratio drag reduction rate reaches a certain asymptotic range. One can easily comprehend from the figure that increasing blowing ratio (BR) has an optimal rate for maximum skin friction drag reduction rate (R), which is independent of input energy. Accordingly, we can elucidate a threshold value for blowing ratio to obtain a maximum saving rate at a fixed Reynolds number.

Subsequently, relationship between Gain (G) to the net energy saving rate (S) is presented through Figure-2.30(b) from the same data. In this figure, saving rate (S) is maximum for highest Blowing ratio (BR) but Gain is minimum. In addition, DNS data from Kametani and Fukagata (2011) is also compared. A similar trend is clearly observable between the experimental results and DNS data. As reasoning for the large deviation found between present experiments to the DNS data, shear Reynolds number difference and idealized assumptions for DNS data can be stated. Therefore, the actual control efficiency presented here should be much less than the ideal values from DNS data.

Ideal value means calculations without considering mechanical losses in the pumping process. It should be noted that the S and G from Kametani and Fukagata (2011) is not as the same definition from the data from the present data. Gain is significantly higher for the former due to the fact that, Kametani and Fukagata (2011) used global friction coefficient, moreover, they have also concluded that control performance indices (G) should be much less when deviated from the ideal values. Instead experimental results are from local skin friction coefficient and input work from unit area. DNS performed in this case ($Re_{\tau,SBL}$ = 160) was way smaller than the experimental data presented here.

## 2.4 Conclusion

We have conducted a series of LDA measurements of 2D profiles along wall normal distance. The measurements were performed in a TBL over smooth surface as reference, later, the TBL was manipulated using wall normal blowing e.g MBT from a finite perforated surface. Blowing was varied with two different velocities namely B1, B2 and B3, where B1 refers to the perforated surface without blowing. Non-dimensional conversion of the blowing velocity ($V_W$) into blowing ratio (BR) expresses the relationship with free stream velocity. Here, B2 varied between 0.76% ~ 0.17% of BR and B3 varied between 1.52% ~ 0.34% of BR respectively.

In light of the above mentioned results presented we can summarize the results of the present paper in the following categories;

(a) Reference SBL data was compared with LES and HWA data from Eitel-Amor (2014) and Örlü and Schlatter (2013) respectively using inner and outer scaling parameters. Present experimental technique was well suited to resolve the near wall region. In addition, diagnostic plot function was also applied in order to compare data independent of the friction velocity. A very good agreement was found between measured data and the reference LES data upto $4^{th}$ order moment.

(b) Wall shear stress was measured using direct measurements of the velocity profiles from viscous sub-layer. Therefore, friction coefficient was obtained and compared with the empirical relationship including other experimental data. This was found to in excellent agreement for SBL reference cases. MBT can reduce friction coefficient significantly and the rate of reduction depends on the blowing ratio. As an example at $Re_{\theta,SBL}$ = 3670, a small blowing ratio of 0.17% can achieve 16% of friction drag reduction. At the same Reynolds number, doubling the blowing ratio can reduce 21%. Therefore, we conclude that increasing the blowing ratio without limit does not change the rate of reduction significantly.

(c) MBT strongly affect the integral properties such as boundary layer thickness, displacement thickness, momentum thickness and shape factor. With increased blowing, integral properties are also increased.

(d) MBT strongly influence first and second order statistical moments of the streamwise and wall normal component of the velocity. An interesting observation is, that due to the wall normal manipulation of the flow, the influence is mainly observable in streamwise velocity component, while wall normal component experiences a smaller impact. Outer scaled profiles of mean streamwise velocity exhibit modification of the profiles within the wall distance $y/\delta = 0.005 \sim 0.55$, Inner scaled mean streamwise profiles show a deviation from the logarithmic law profile. The magnitude of the deviation depends on the blowing ratio and this is observed at all Reynolds number. Streamwise velocity fluctuation exhibit an enhanced outer peak and this peak is even more pronounced when blowing ratio is increased. For the wall normal velocity component, the mean shows stronger positive deviation at the log layer, while the velocity fluctuations are enhanced close to the wall and finally, collapses to the SBL reference at $400 \leq y^+ \leq 600$ e.g within log layer.

(e) We have calculated the performance indicator parameters using friction coefficients at different blowing ratio. For increasing blowing ratio, drag reduction rate reaches a certain asymptotic range. This proposes a procedure to calculate performance indicators (R, S and G) for experimental cases where local blowing is applied. For further experiments this procedure will effectively offer a guideline to select an optimized rate of blowing.

Advantage of the present measurement is the high spatial resolution of the statistical profiles, especially in the vicinity of a wall. However, it is not possible to analyse the behaviour of LSM and VLSM from the point based measured data. To do so, spatial measurements such as PIV are required, while this sacrifices the high spatial resolution given by the LDA. Therefore, present paper has been written focusing in to the statistical perspective of a turbulent boundary layer under the influence of blowing.

# 3 High Reynolds number experiment

## 3.1 Introduction

Within the context of this chapter, measurements at high Reynolds number will be presented. Here, effect of uniform blowing has been experimentally investigated in a zero pressure gradient TBL. Laboratoire de Mécanique des Fluides de Lille wind tunnel used for the measurement was particularly suitable to obtain high resolution data (Boundary layer thickness, $\delta > 0.24m$) at high Reynolds number with Stereo Particle Image Velocimetry (SPIV) measurements. The data presented in this chapter covers a large range of high Reynolds number flow e.g. $Re_\theta = 7500 \sim 19763$ where Reynolds number is based on the momentum thickness. Upstream effect of blowing was varied from 1% ~ 6% of free stream velocity by tuning the flow rate of the compressed air and measurements were taken downstream after a short interval. In order to access statistics and turbulence properties of the TBL with focus on the logarithmic and outer region, the streamwise SPIV plane (Vertical plane parallel to flow direction) configuration was used to obtain velocity fields. The results presented here has already been published in Hasanuzzaman et al. (2020).

## 3.2 Experimental procedure

In this section experimental facility used to obtain data will be discussed followed by the description of the measurement method itself.

In order to indicate different locations inside the wind tunnel, Cartesian co-ordinate system x, y and z direction is used to indicate streamwise (longitudinal), wall normal (vertical) and spanwise (transverse) direction respectively. In all cases streamwise direction on the surface of the present boundary layer was measured from the tripping location. Here, positive distance along x, y and z axis indicate further away from tripping, wall and wall centerline respectively. Reported boundary layer develops over the lower flat wall where an artificial tripping is installed exactly at the leading edge. Tripping is done with a 4 mm high spanwise metal bar followed by a grade 40 sand paper spacing $0.093m \times 2m$ in streamwise and spanwise direction respectively which facilitates direct transition to turbulent boundary layer. Therefore, with the help of tripping at the inlet, a thick boundary layer develops in the order of 0.24 m at a streamwise distance of $X = 19.2m$ from tripping.

Present experiment was conducted in zero pressure gradient condition e.g. difference between mean pressure gradient along streamwise direction was close to zero or

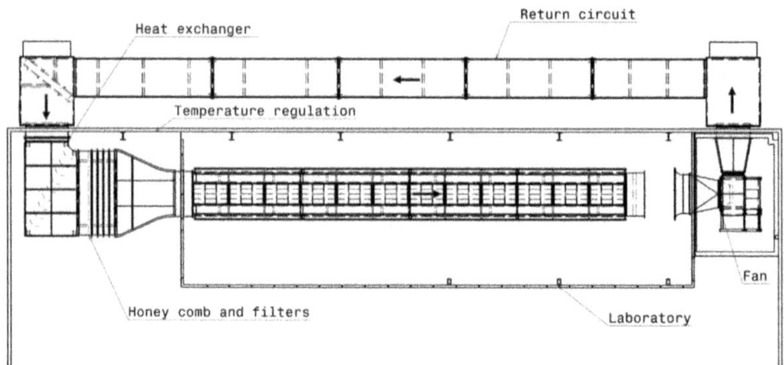


**Figure 3.1:** Schematic of the top view of wind tunnel (Cuvier et al. (2017)). Primary flow path is indicated with arrows.

Zero Pressure Gradient (ZPG) condition was assumed to ease the experimental condition. Moreover, this SBL condition also exhibit the analogous characteristics to different length scales of structures present to the application range. For this experiment, idealized environment was assumed using kinematic and geometric similarity. Stronger effort was directed towards the data acquisition in near wall region. Another particular characteristics which is often neglected in such experiments is a steady outer flow required with low free stream turbulence level, which is in present case was lower than 0.25% than that of free stream values.

Only uniform blowing was applied in SBL cases as a mean to flow control. For reference SBL data boundary conditions at wall are as similar to the one as "no-slip" conditions e.g $[\bar{u}_x, \bar{u}_y, \bar{u}_z]_{wall} = (0, 0, 0)$ and at the outer edge of the boundary layer $[\bar{u}_x, \bar{u}_y, \bar{u}_z]_\infty = (U_\infty, 0, 0)$. Here, $\bar{u}_x$, $\bar{u}_y$ and $\bar{u}_z$ are the streamwise, wall normal and spanwise component of velocity averaged in time respectively and $U_\infty$ refers to the free stream velocity at the outer edge. Therefore, special consideration should be taken in order to interpret Reynolds number with subscript "SBL" as stated earlier. Such for uniform blowing measurements, boundary condition at the wall is, $[\bar{u}_x, \bar{u}_y, \bar{u}_z]_{wall} = (0, V_w, 0)$, here, $V_w$ is the velocity of blowing applied in perpendicular direction coming from wall. Details of this blowing air parameter will be discussed in Section-3.2.2.

Within the scope of present literature, results are presented for $7495 \leq Re_{\theta,SBL} \leq 18094$, here, $Re_{\theta,SBL} = U_\infty \theta / \nu$ (with the momentum thickness ($\theta$) and the free stream velocity ($U_\infty$)). The following flow condition is comparable to $2186 \leq Re_{\tau,SBL} \leq 5482$ using viscous parameter where $Re_{\tau,SBL} = \delta u_\tau / \nu$ which is also known as $\delta^+$. Another important characteristics that is strongly influenced by the control technique is the shape factor ($H = \delta^* / \theta$) where $\delta^*$ is the displacement thickness. This parameter indicate not only the turbulent state of the boundary layer but also the modifications made to the mean properties when control is applied.

### 3.2.1 The wind tunnel facility

All measurements were carried out in the large boundary layer wind tunnel of "Laboratoire de Mécanique des Fluides de Lille – Kampé de Fériet (LMFL). This facility is particularly suitable for high resolution measurements at high Reynolds numbers of SBL. The wind tunnel used for this experiment has a closed loop configuration which is particularly suitable for non-intrusive optical measurements.

The test section of the wind tunnel is $20.6m$ long with a cross section of $2m^2$ with vertical and transverse lengths of 1 m and 2 m, respectively. As the test section has an optical access from all sides along the complete length of it, high quality PIV measurement through the optical access is possible. Figure-3.1 present the sketch of the wind tunnel drawn based on the top view of it. Incoming air to the plenum chamber is passing through an air-water heat ex-changer in order to provide an iso-thermal flow where efficiency is kept within $\pm 0.15^{\circ}C$. Subsequently, air through the guide vanes undergoes a relaminarization process via honeycomb screens and grids. Thereafter, contraction takes place with a ratio of 5.4 : 1.

Free stream/external velocity ($U_\infty$) can be regulated in the range of $3 \sim 9m/s$ at the entrance of the test section with a precision of $\pm 0.5\%$. The free stream turbulence is below 0.2 %. This wind tunnel allows us to investigate a wide range of local Reynolds number range based on local momentum thickness.

### 3.2.2 Uniform blowing setup and characterization

Uniform vertical blowing was provided with a perforated/blowing surface from the beginning of the test section at $x_1 = 18.424m$ and ends at a streamwise distance of $x_2 = 18.845m$. Figure-3.2(a) and Figure-3.2(b) displays the parallel and top view projection of the experimental segment of wind tunnel in 2D space respectively. In both cases, region of the wall from which uniform blowing was applied is indicated with pink region. Perforated surface is $0.55m$ and $0.42m$ in width and length respectively. This was placed symmetrically at spanwise center of the test section. Although, lower wall of the test section is composed of different segments, careful effort was provided to keep the streamwise alignment better than $0.1mm$.

Perforated surface is constructed from a $20mm$ thick stainless steel plate where 4514 holes were precisely drilled in staggered arrangement[1]. From Figure-3.2(c) exhibit the arrangement of the holes in top view to the wall frame of reference. Each holes were precisely drilled with a diameter of $3.6mm$, were organized from each other keeping their center $14.4mm$ in streamwise and $7.2mm$ in spanwise direction.

[1]Design of the holes arrangement was adopted from the Tailored Skin Single Duct (TSSD) design from Horn et al. (2015), this was discussed in detail from Krishnan et al. (2017). The original micro-perforated surface was designed for A340-300 with a hole diameter to to spanwise distance ratio of 1:2. Scale modification was done based on $\delta$ for spatially developped thick SBL condition such as LMFL wind tunnel.

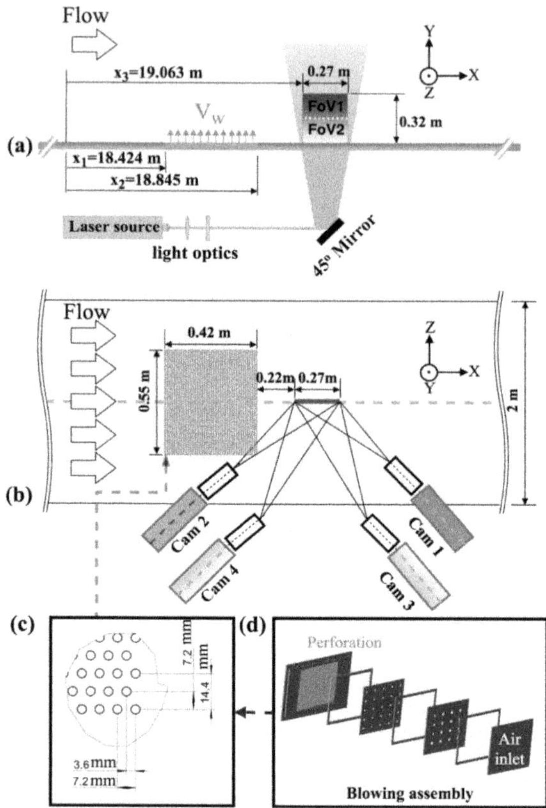


**Figure 3.2:** (a) Parallel view scheme of uniform blowing experiment over flat plate SBL showing the laser light sheet and the measured field of view of the streamwise plane, location of uniform blowing is marked with the pink region, (b) Top view schematic of the same (c) Drawings showing the arrangement of holes and (d) Break down of different segments of blowing assembly

Perforated surface is comprised of different components attached together in order to form a box shaped device where air is remained sealed and can only be transmissible from inlet all the way through the perforated region. Therefore, the blowing device formed out of different sections and will be termed as "Blowing assembly" hereafter. Figure-3.2(d) displays isometric view of different segments where two additional surfaces between the air inlet and perforation was used along with other additional fixtures as such that the inlet air pressure is uniformly distributed inside the "Blowing assembly". Here, region marked with shaded pink indicate the opposite side of the perforated surface. Pneumatic sealing was used between each fixtures to prevent the air loss and was checked in every stage while assembling together. Blowing assembly was connected with four poly-amide air tubes equipped with Legris fast connectors through which dry compressed air was provided. Pressurized air was regulated with a pressure regulator and valve. The mass flow rate was measured with an appropriate vortex flow meter(accuracy $\pm 1.5\%$ on volume flow rate) and a temperature and pressure sensor to get the density (accuracy $\pm 1\%$). Total uncertainty on the imposed flow rate to get the desired velocity through the perforated wall is then $\pm 2\%$.

In order to quantify the magnitude of blowing, blowing fraction ($F$) will be used hereafter. This is simply the ratio in percentile between blowing velocity ($V_w$) and free stream velocity at the entrance of the test section. Therefore one can derive the formula for the blowing fraction as $F = (V_w/U_\infty) \times 100$. The blowing velocity ($V_w$) is determined by the total mass flux by the compressor and the sum of the perforated hole surfaces. For each Reynolds number being investigated, upstream blowing was varied in 4 different ratios, namely 0%, 1%, 3% and 6%. For each Reynolds number and blowing ratio, each data set will be termed as cases corresponding to the different planes of measurement. Air flow rate for different blowing ratio was varied between $2 \sim 500 m^3/hour$.

One of the primary objective of the present experiment was to investigate the effect of maximum blowing ratio which would be sustainable to keep the TBL profile before reaching into the potential layer (free stream). Maximum blowing ratio of 6% was selected as the maximum limit as it is already a strong mass flow rate injected compared to the one of the viscous sublayer (about 50 times at maximum velocity). The aim of the study is to find alternative flow control strategy so the energy injected should be as small as possible to get positive balance.

Measurements from a smooth surface SBL was also obtained in addition to the different ratios of blowing. Analysis of this data from upstream smooth wall will be used for wind tunnel characterization and to observe the changes in SBL mean properties compared to the different ratios of upstream blowing.

### 3.2.3 Particle image velocimetry (PIV)

In order to obtain all three components of the velocity, SPIV technique was used. The flow was successively measured in XY plane using SPIV arrangement. Description of this set-up will be discussed in the following section.

The velocity profiles downstream to the uniform blowing region acquired in a SPIV plane started after $0.22m$ from the end of the blowing region. Distance of this plane relative to the start of the test section is $19.063m$ downstream. The Field of View (FoV) as indicated with shaded blue in Figure-3.2(a) was $0.27m \times 0.32m$ in streamwise and wall normal direction respectively.

Laser light sheet was produced from a BMI laser with $200mJ/pulse$ through the bottom glass surface which is shown in Figure-3.2(a). Incoming laser beam was passing through a spherical ($f' = 5.6m$) lens placed at 0.5 m and a cylindrical ($f' = -0.25m$)

lens placed at 0.75 m from the laser source outlet respectively before directed by a 45° mirror placed below the glass wall and 8.3 m downstream from the laser output (distance from beam exit to bottom glass is 9 m). Therefore, creating a light sheet with uniform thickness[2] of 0.6 mm as indicated with green in Figure-3.2(a). Two cavities of the pulsed laser were synchronized with the camera at a frequency $f_{acq}$ = 4 Hz.

The SPIV set-up was consisting of 2 separate stereo systems placed on top of the other, this can be comprehended from Figure-3.3(a). Camera 1 and 2 (upper part of FoV), camera 3 and 4 (see Figure-3.2) will be termed as SPIV system 1 and 2 respectively for the subsequent discussion. Here, 4 sCMOS camera were used for this setup, each of the camera CMOS sensor array having streamwise and wall normal resolution of 2560 × 2160 pixel$^2$ with a pixel size of $6.5\mu m$. Larger pixels of each camera were aligned with the streamwise axis of the flow field i.e. larger side of the camera sensor was imaging streamwise extent of FoV. Each camera lenses were mounted with a $135mm$ Nikkor lens and was set at $f\#$ = 8. The camera arrangement for XY plane is displayed in Figure-3.2(b), 2 cameras were arranged in angular configuration within each stereo system following the description from Prasad (2000). Complete FoV for XY plane was obtained with the overlapping of two planes acquired from each stereo system on top of the other where camera 1 and camera 2 was responsible for the top plane (FoV 1 in Figure-3.2(a)), Camera 3 and camera 4 was responsible for the bottom plane (FoV 2 in Figure-3.2(a)). Therefore, a common region was present between each field in the order of $10mm$ at $y$ direction. In Figure-3.2(a), common region is indicated with the dotted yellow line. Cameras were mounted on a custom made bench with a Scheimpflug adapter, approximately $0.13m$ away from the test section glass wall (working distance of about 1.7 m). As summarized in the Table-3.1, magnification value was 0.083 in order to resolve most of the structures present in the logarithmic and outer region of the flow. With this value, the stretching factor of the stereo viewing angle (about 45deg) and the approximate value of the external velocity, a $\Delta t$ was computed and impose to obtain about this 12 pixels displacement in the external region. Before recording, some random snapshots were taken and analysed rapidly in 2D2C to check that the correlation was working well with this 12 pixels dynamics. With these tests, we have optimised the $\Delta t$ around the starting value to get high dynamic and good correlation (low number of spurious vector). The final $\Delta t$ used was leading to 12 pixels displacement in the external region of the boundary layer with the 2D2C tests before recording (small correction compared to the first estimation).

Two cavities of the BMI laser was triggered with varying time delay of $\Delta t = 135 \sim 405 \mu s$ depending on the Reynolds number of the flow. The incoming laser was applied on the XY plane as shown in Figure-3.3(b). Acquisition frequency ($f_{acq}$) was set at a constant value of 4 Hz for each data set. A total of 3000 velocity fields were obtained with all

[2] The laser sheet thickness was computed with laser beam propagation formula (non Gaussian beam with $M^2$ = 1.2). With the optic used, the half-angle divergence of a Gaussian laser beam (theta) was computed and the light sheet thickness which is then equal to the beam waist diameter 2 × lambda × $M^2/(\pi \times$ theta), with the laser wavelength, lambda = 532 nm. In order to confirm, the result was checked by making some light sheet impact on special paper, therefore, thickness of the light sheet was measured with binocular magnifier with an accuracy of ±0.1 mm.

three components of velocity where w (Z direction) component is the out of plane motion (Table-3.1).

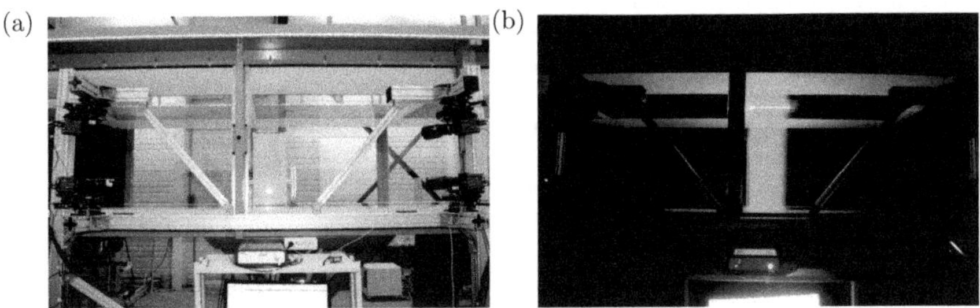

**Figure 3.3:** Photograph of (a) SPIV camera arrangement at LMFL wind tunnel during calibration and (b) Incoming laser beam direction from the bottom surface.

The flow was seeded with tracer particles by a Hazebase Base Classic fog machine located in the wind tunnel diffuser section. Particles are globally provided by a repeating evaporation and condensation process using a solution of poly-ethylene-glycol and water. The mean particle diameter is approximately 1 $\mu m$ and have a lifetime of around 10 minutes circulated within the closed circuit wind tunnel.

A nomenclature stated in Table-3.1 indicate in a sequence the reference plane of measurement, free stream velocity, surface condition and rate of blowing fraction applied. Where, first 2 letters in capital 'XY' indicate reference plane of measurement from streamwise-wall normal plane, third letter 'U' followed by a fourth or fifth digit indicate free stream velocity and the fourth letter 'S' or 'F' indicate surface condition which is 'SBL' or blowing respectively. Finally, the last digit indicate the rate of blowing fraction applied. PIV images were processed using an in house (LMFL) modified version of MatPIV code. Image deformation due to stereoscopic aberration was adjusted using the Soloff back projection/re-construction of three dimensional warping technique (Soloff et al. (1997)). This is the same 3D warping technique described in Coudert and Schon (2001) and implemented in the calibration process described as "Self-calibration" applied in Wieneke (2005). Laser light sheet misalignment correction was done by cross-correlating the two PIV mapped images, which gives us the opportunity to correct the error on the 3D vector origin.

During calibration, a precision translation stage was used to offset along the corresponding calibration axis (z-axis). For the measurement in XY plane, the translation stage was moved in the spanwise direction for $-0.005m \leq z \leq 0.005m$ where each step was $\Delta z = 0.0005$. A sequence of 21 images for the calibration target were acquired on both stereo systems. Here, 0.8 m $\times$ 0.34 m rectangular calibration target was printed with dark crosses with a precision stretching of gap between each circles of 0.00916667 m and 0.0091591 m along x and y-axis respectively. Therefore, high degree of sharpness on the camera sensor was attained by applying optimal aperture ($f\# \geq 8$)

of the mounted lenses and careful application of the Scheimpflug condition using the Scheimpflug adapters following Prasad and Jensen (1995).

The final Interrogation Window (IW) size was 18 x 24 (corresponding to 1.9 mm x 1.9 mm or about 45 wall-units at the highest Reynolds number). Final IW varied between $15.3^+ \sim 43.2^+$ depending on the Reynolds number of the flow (see Table-3.1).

A mesh of grid points were created in the physical space and were then projected into the camera space thanks to the Soloff function (Soloff et al. (1997)). The displacement is then evaluated on each camera on these projected grid which are not regular. Finally, Soloff reconstruction method was used to obtain the 3C velocity fields. Common grid points from the calibration target is then reconstructed using 2D2C component. Number of mesh points for top stereo system had a 334 points along x-axis and 206 points along y-axis with an increment of 0.0008 m along both the axis (corresponding to 7.435 pixel along x-axis or 9.682 pixel along y-axis). Therefore, recombination of both stereo systems provided a mesh steps of 334 and 402 in x-axis and y-axis respectively where, first mesh point along y-axis was determined at 0.0012 m or 14.5 pixels from the wall.

### 3.2.4 Measurement uncertainty

As discussed in sub-section-3.2.3, SPIV system-1 and system-2 had grid points (measurement point in the object plane) $[N_{x,sys1}, N_{y,sys1}]$, $[N_{x,sys2}, N_{y,sys2}]$=[334 , 206], [334, 206]. Merging of the two overlapping system took place at the intersection of the both systems and causes an overlapping of 334 and 10 vector points in streamwise and wall normal direction.

Thus merging region extends in 10 and 333 points along wall normal and streamwise directions respectively, e.g. number of grid points along X and Y axis for the merging region would be $[N_{x,m}, N_{y,m}] = [333, 10]$. Merging of two separate Stereo systems is done by simply taking the average of each component from two 2D3D velocity fields. Therefore, SPIV bias error and random PIV uncertainty can be estimated following the procedure proposed by Kostas et al. (2005), Herpin et al. (2008) and Cuvier (2012).

Total uncertainty on the mean velocity components of a 2D3C field can be calculated using the velocity data for two separate SPIV systems within the merging region. Equation-3.1 is used to determine the bias error or the PIV uncertainty. Here, $u_{sys1}$ and $u_{sys2}$ represent the streamwise velocity component for SPIV system 1 and 2 respectively. Similarly, bias error for wall normal (v) and spanwise (w) component of the velocity field was obtained using difference between simultaneous image pairs of two systems in the merging region.

The random PIV uncertainty ($\sigma_{\epsilon_u}$) with 95% confidence interval for the streamwise velocity component is determined using Equation-3.2. This equation is also applicable in determining the random error for wall normal ($\sigma_{\epsilon_v}$) and spanwise ($\sigma_{\epsilon_w}$) component of the velocity.

| Plane (1-2) | | XY | |
|---|---|---|---|
| Stereoscopic angle | | $45^\circ$ | |
| Focal Length (mm) | | 135 | |
| Laser sheet thickness (mm) | | 0.6 | |
| Magnification | | 0.083 | |
| Lens aperture ($f\#$) | | 8 | |
| Lens working distance from FoV (m) | | 1.7 | |
| FoV (m × m) | | 0.27 × 0.32 | |
| CMOS array [x × y] (px × px) | | 2560 × 2160 | |
| Overlap region in 'y' (mm) | | 10 | |
| Image acquisition frequency (Hz) | | 4 | |
| IW size [mm × mm] | | 1.9 × 1.9 | |
| No. of records | | $3\times10^3$ | |
| SBL Reynolds number ($Re_{\theta,SBL}$) | 7500 | 12500 | 18100 |
| IW size [$L_{IW}$]($y^+ \times y^+$) | 15.3×15.3 | 29.7×29.7 | 43.2×43.2 |
| FoV [$S_{x,1}$,$S_{y,1}$] | $1\delta$, $1.2\delta$ | $1.1\delta$, $1.32\delta$ | $1.1\delta$,$1.33\delta$ |
| Mesh step [$\Delta_i/l^+_{SBL}$] ($y^+$) | $6.5^+$ | $12.5^+$ | $18.5^+$ |
| Integral time scale [$\Lambda$] (s) | 0.081 | 0.036 | 0.024 |
| Convergence uncertainty on $\bar{u}$, | | | |
| - SBL [$U_{\bar{u},SBL}$](%) | ±1.1 | ±0.7 | ±0.7 |
| - at F=0% [$U_{\bar{u},F0}$](%) | ±1.1 | ±0.8 | ±0.7 |
| - at F=1% [$U_{\bar{u},F1}$](%) | ±1.2 | ±0.8 | ±0.7 |
| - at F=3% [$U_{\bar{u},F3}$](%) | ±1.2 | ±0.9 | ±0.8 |
| - at F=6% [$U_{\bar{u},F6}$](%) | ±1.4 | ±1.1 | ±1.1 |
| Convergence uncertainty on $\sqrt{\overline{u'^2}}$, [$U_{u_{rms}}$](%) | | 5.1 | |
| Rec. | | | |
| - SBL | XYU3S | XYU6S | XYU10S |
| - at F=0% | XYU3F0 | XYU6F0 | XYU10F0 |
| - at F=1% | XYU3F1 | XYU6F1 | XYU10F1 |
| - at F=3% | XYU3F3 | XYU6F3 | XYU10F3 |
| - at F=6% | XYU3F6 | XYU6F6 | XYU10F6 |

**Table 3.1:** PIV recording parameters

Table-3.2 compiles the error estimated at streamwise and wall normal distances of [X/$\delta$, Y/$\delta$, $Re_{\theta,SBL}$] = [70.41,0.58, 7495], [78.46, 0.64, 12542] and [79.14, 0.65, 18094]. Here, bias and merging error for all three component is compiled as a fraction of $U_\infty$. In table-3.2, $Re_{\theta,SBL}$ and Rec. indicate the reference SBL case and corresponding labelling respectively similar to the Table-3.1.

$$\Delta u = \pm\langle u_{sys1} - u_{sys2}\rangle \tag{3.1}$$

$$\sigma_{\epsilon_u} = \pm(u_{sys1} - u_{sys2})_{RMS} \tag{3.2}$$

| | | Bias error (%U$_\infty$) | | | Random error (%U$_\infty$) | | |
|---|---|---|---|---|---|---|---|
| $Re_{\theta,SBL}$ | Rec. | -u | -v | -w | u | v | w |
| 7500 | XYU3S | 0.03 | 0.03 | 0.07 | 0.6 | 0.5 | 0.8 |
| 12500 | XYU6S | 0.04 | 0.03 | 0.06 | 0.7 | 0.6 | 0.9 |
| 18100 | XYU9S | 0.03 | 0.04 | 0.06 | 0.9 | 0.5 | 1.1 |

**Table 3.2:** SPIV error analysis using the overlapping region. The errors are given in percentage of $U_\infty$.

### 3.2.5 Convergence of data

PIV images were acquired as such that statistically independent measurement at 4 Hz were taken for each recordings. Although, there remains a question regarding statistical convergence of the data. It is therefore, important to obtain sufficient amount of ensembles in order to have acceptable convergence. For turbulent flows such as TBL, large gradient of the mean streamwise velocity is present from the region of high frequency fluctuations to the regions with lower frequency fluctuations. Kähler et al. (2016), has showed the convergence of moments from the streamwise fluctuation for different sampling rate. They showed that, data points in the free stream region converges faster than the near wall regions. In addition, they recommend that atleast 1000 samples through time are required to obtain the first and second order statistics correctly. Therefore, it necessitates a convergence study for the data presented here.

Dixon and Massey (1957) suggested that statistical convergence uncertainty on a sample mean and standard deviation with 95% confidence interval (and Gaussian distribution) can be calculated using Equation-3.3 and Equation-3.4 respectively. Here, both equations are derived for the streamwise velocity component. Similar expressions can also be derived for wall-normal and spanwise velocity component. $N_{eff}$ is the total number of time steps or total number of samples acquired with statistical independence. In addition, Ahn and Fessler (2003) suggested that error rate for the following formulae are less than 1% when no of samples$\geq$ 30. In order to confirm that the samples were uncorrelated in time, time separation between two samples were selected as such that the sampling rate was less than 1/2×integral time scale as suggested by Benedict and Gould (1996), provided that the total acquisition time was fixed. Here, integral time scale was calculated with $\Lambda = \delta/U_\infty$. In other words, sampling rate was 3 to 10 times larger than that of the integral time scale depending on the Reynolds numbers. In the present experiment, total number of samples acquired e.g $N_{eff}$=3000 at acquisition frequency of 4 Hz which corresponds to sampling rate more than $2\Lambda$. Convergence uncertainties on the mean and turbulence intensity from streamwise velocity component is therefore, presented in Table-3.1.

| $U_\infty$ (m/s) | $u_\tau$ (m/s) | $\nu/u_\tau$ ($\mu/s$) | $C_f$ [$\times10^3$] | $\delta$ (m) | $\delta^*$ (m) | $\theta$ (m) | H - | $Re_x$ [$\times10^6$] | $Re_\theta$ [$\times10^3$] | $Re_{\delta^*}$ [$\times10^3$] | $Re_\tau$ - |
|---|---|---|---|---|---|---|---|---|---|---|---|
| 3.4 | 0.121 | 124 | 2.53 | 0.271 | 0.045 | 0.0331 | 1.36 | 4.33 | 7.5 | 10.2 | 2186 |
| 6.78 | 0.2345 | 64 | 2.4 | 0.243 | 0.037 | 0.0277 | 1.32 | 8.64 | 12.5 | 16.5 | 3800 |
| 10.14 | 0.3455 | 44 | 2.32 | 0.241 | 0.0348 | 0.027 | 1.3 | 12.93 | 18.1 | 23.5 | 5482 |

**Table 3.3:** Mean properties (for reference cases) of the SBL in LMFL wind tunnel at X = 19.2 m obtained from SPIV measurement.

$$U_{\bar{u}} = \frac{\Delta\bar{u}}{\bar{u}} = \pm\frac{1.96}{\sqrt{N_{eff}}} \times \frac{\sqrt{\bar{u'^2}}}{\bar{u}} \tag{3.3}$$

$$U_{u_{rms}} = \frac{\Delta\sqrt{\bar{u'^2}}}{\sqrt{\bar{u'^2}}} = \pm1.96\sqrt{\frac{2}{N_{eff}}} \tag{3.4}$$

### 3.2.6 Boundary Layer Characterization

A priori to the control application using uniform blowing, experimental facility was characterized using conventional smooth surface. Characterization was done using the similar SPIV setup described earlier except wall shear determination. Recently, Willert (2015) conducted high magnification PIV experiments in the same facility where very accurate determination of wall shear was possible. In the present experiment, wall shear was obtained using the same method for the recording XYU3S, XYU6S and XYU10S (Foucaut et al. (2018)). These data sets represent the characterization parameters for *SBL* condition with smooth surfaces. Table-3.3 summarized the mean properties of the SBL reference cases. Later mean properties with the application of MBT is also determined using the same streamwise stereo PIV set-up, except wall shear which was not determined. No additional suffix was used for the reference cases as normalization to these cases are done using respective inner and outer parameters.

### 3.2.7 Validation

### 3.2.8 SBL mean properties

As the size of the field of view is only of the order of one boundary layer thickness, the streamwise direction can be taken as homogeneous direction following an approach

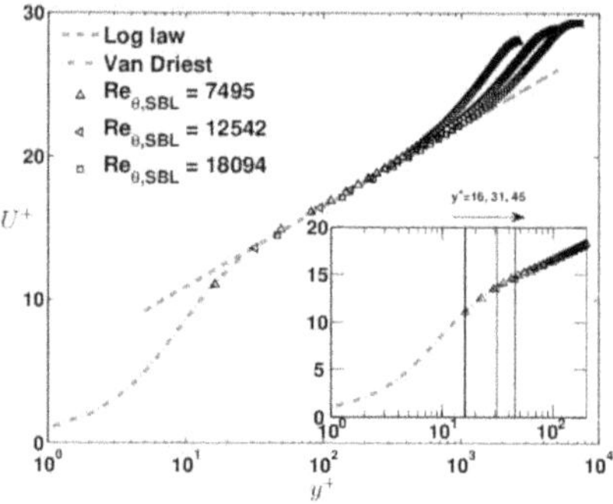


**Figure 3.4:** SPIV data from XY plane for reference *SBL* cases. Mean streamwise velocity in wall units ($U^+$) plotted against dimensionless wall units ($y^+$). In the inset figure, same data is plotted in order to highlight the merging of the experimental data to the Van Driest profile within near wall region. One out of every five data point are presented for clarity.

law $\delta/x = 0.37Re_x^{-0.2}$ from Schlichting (1960). Therefore, data presented for mean statistics is then averaged in both time and space as the FoV in streamwise direction was less than the boundary layer thickness in streamwise direction.

Mean streamwise velocity and corresponding wall normal distance is expressed as $U^+$ and $y^+$ in Equation-3.6. Literature suggestion regarding the extent of these layers are, viscous sub-layer: $0 \leq y^+ \leq 3-5$, buffer layer: $3-5 \leq y^+ \leq 30$ and log layer: $30 \leq y^+ \leq 0.1\delta^+$. Although, there are several arguments regarding the exact determination of these layers, but this paper only deals with the effect of blowing.

$$P = -\overline{u'v'}(\frac{\partial U^+}{\partial y^+}) \tag{3.5}$$

$$U^+ = \frac{1}{\kappa}\ln(y^+) + C \quad \text{for } 30 \leq y^+ \leq 0.1\delta^+ \tag{3.6}$$

where,

$$U^+ = \overline{u}/u_\tau$$
$$y^+ = yu_\tau/\nu$$

$$\Xi = \left(y^+\frac{dU^+}{dy^+}\right)^{-1} \tag{3.7}$$

In order to show the quality of the XY stereo PIV plane to extract mean and turbulent profiles, first, in Figure-3.4, mean streamwise velocity profiles of SBL cases are drawn along with Van Driest profile following Equation-1.30 (Van Driest (1956)) and logarithmic profile following Equation-3.6. Inset figure is given in order to enlarge the near wall region and a good merging of the experimental data to the Van Driest profile can be observed. First data for $Re_{\theta,SBL}$ = 7495, 12542 and 18095 were obtained at $y^+$ = 16, 31 and 45 respectively. At sufficient spatial resolution, SBL profiles at all three Re displays distinct logarithmic and wake region. In general, logarithmic region starts at $y^+ \approx 30$ and ends at ≈ 400, 700 and 1000 from lowest to highest Re.

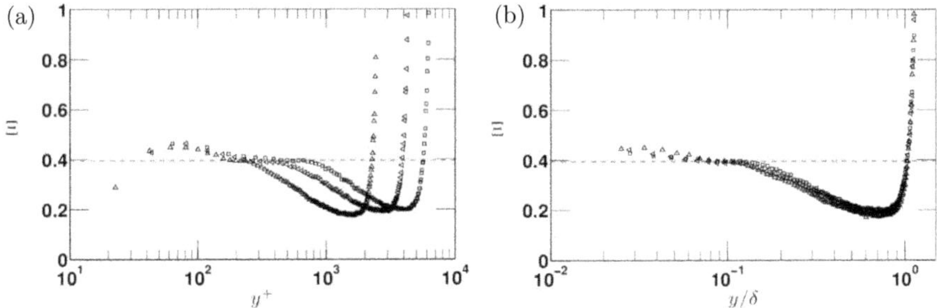


**Figure 3.5:** Normalized slope of mean profile Ξ, shown in (a) inner scaling, (b) outer scaling where, △, ◁, and □ presents SBL data at $Re_{\theta,SBL}$ = 7500, 12500, 18100 respectively.

In Equation-1.30, $A^+ = 26$, the value of Von Karman constant ($\kappa$) and C is taken as 0.40 and 5.3 respectively. $\kappa$ was calculated using Equation-3.7, thereafter plotting Ξ, $\kappa$ was determined following the plateau of the log layer. Boundary layer characteristics of SBL cases are reported in Table-3.1. Interested readers are advised Foucaut et al. (2018) for details of the measurement technique.

Figure-3.6 shows the turbulence intensity of u, v and w components scaled with the inner parameters and compared to the well resolved LES data at $Re_{\theta,SBL}$ = 7603 from Eitel-Amor (2014). In order to obtain clarity for the experimental data in the figure, one out of every five data points are plotted here. Both LES and experimental data shows excellent agreement except the first data from experiment. Reflection from the wall is very common problem observed for PIV experiments.

Figure-3.7 shows the profiles of turbulence intensities for all three components obtained for all SBL cases (at $Re_{\theta,SBL}$ = 7500, 12500 and 18100). Here, markers represent experimental data obtained with SPIV measurements and continuous line represents the LES data from Eitel-Amor (2014). One out of every ten data is plotted in the following figure in order to have better clarity except the enlarged figure. Turbulence intensity profiles for three Reynolds numbers are scaled in wall units and as a function of both $y/\delta$ and away from the wall. Inset figure is plotted for the same data as a function of $y^+$ in order to enlarge the near wall region. Very good universality is observed for both representations along with the numerical data. Although a certain deduction regarding

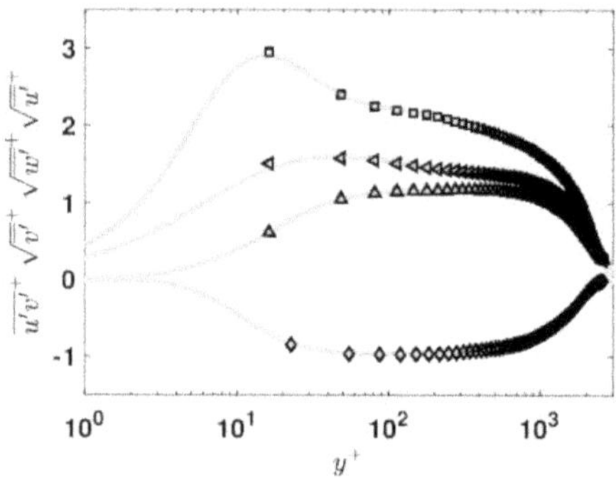


**Figure 3.6:** Profiles of RMS values (turbulence intensities) of velocity fluctuations and RSS obtained. SPIV data (markers) is compared with LES data (continuous line). ——, LES data at $\mathrm{Re}_\theta = 7603$ from Eitel-Amor (2014) and markers represent present SPIV data at $\mathrm{Re}_\theta = 7495$: □, $\sqrt{\overline{u'^2}}/u_\tau$;△, $\sqrt{\overline{v'^2}}/u_\tau$; ◁, $\sqrt{\overline{w'^2}}/u_\tau$ and ⋄, $\overline{u'v'}/u_\tau^2$.

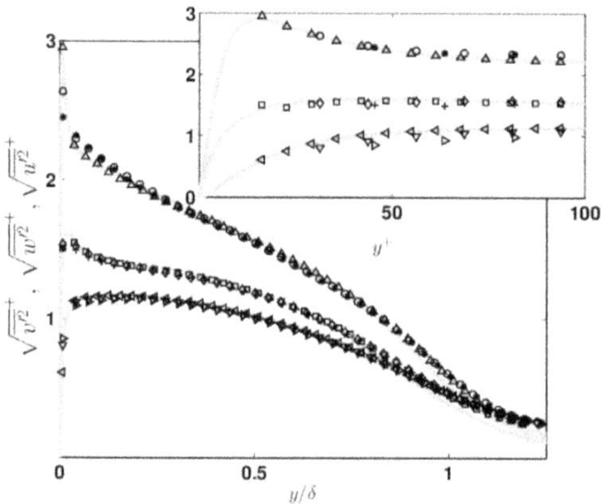


**Figure 3.7:** Profiles of turbulence intensities: ——, LES data at $\mathrm{Re}_\theta = 7603$ from Eitel-Amor (2014) and symbols represent the present SPIV data in XY plane; $\sqrt{\overline{u'}}^+$, $\sqrt{\overline{v'}}^+$ and $\sqrt{\overline{w'}}^+$ are presented for all SBL cases respectively with: △, ◁, and □, at $\mathrm{Re}_{\theta,SBL} = 7500$; ∘, ▽ and ⋄ at $\mathrm{Re}_{\theta,SBL} = 12500$; ∗, ▷ and + at $\mathrm{Re}_{\theta,SBL} = 18100$.

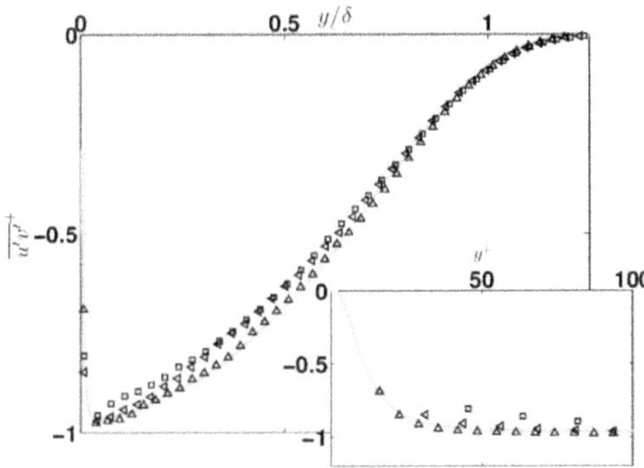


**Figure 3.8:** Profiles of RSS $(\overline{u'v'}^+)$: ——, LES data at $\mathrm{Re}_\theta = 7603$ from Eitel-Amor (2014) and symbols represent the present SPIV data in XY plane; $\triangle$, $\mathrm{Re}_{\theta,SBL} = 7500$; $\triangleleft$, $\mathrm{Re}_{\theta,SBL} = 12500$ and $\square$, $\mathrm{Re}_{\theta,SBL} = 18100$;

near wall peak value is not possible due to lack of data from SPIV which is solved in the paper of Foucaut et al. (2018) by combining the result with high magnification PIV.

Figure-3.8 shows the profiles of RSS normalized with inner velocity $(\overline{u'v'}/u_\tau^2)$. This is obtained for the same data and compared to the same LES data as stated before. Zagarola and Smits (1998) suggested that the outer scaled wall location to the inner scaled RSS exhibit a better Reynolds number independence atleast in the outer region, therefore, wall locations are normalized with the corresponding outer length scale e.g $\delta$. In addition, inner scaled representation of the data is presented for the region $0 \leq y^+ \leq 100$. Although Reynolds stresses are being sensitive parameter when compared in a wide range of Reynolds number measurements. A slight variation in the near wall region can also be related to the increasing filtering with Reynolds number. But numerical data agrees well to the corresponding nearest data set ($\mathrm{Re}_{\theta,SBL} = 7495$). In particular, near wall presentation as a function of $y^+$ exhibit better universality than the $y/\delta$ presentation. Figure-3.9 (a) plots the turbulence production (Equation-3.5) along inner scaled wall normal locations in logarithmic abscissa. LES data from Eitel-Amor (2014) is compared with all SBL data at three Re investigated. All data in the plot collapse well to each other. This inner scaling of the production confirmed that it is a near wall effect. Here, Figure-3.9 (b) is presented as pre-multiplied with dimensionless wall distance (suggested by Marusic et al. (2010a)) to dimensionless wall distance, in order to represent the equal areas which shows equal contributions to the bulk production term. For each Reynolds number investigated, all production profiles display a clear peak in the outer region and agrees well to the numerical data for the lowest Reynolds number.

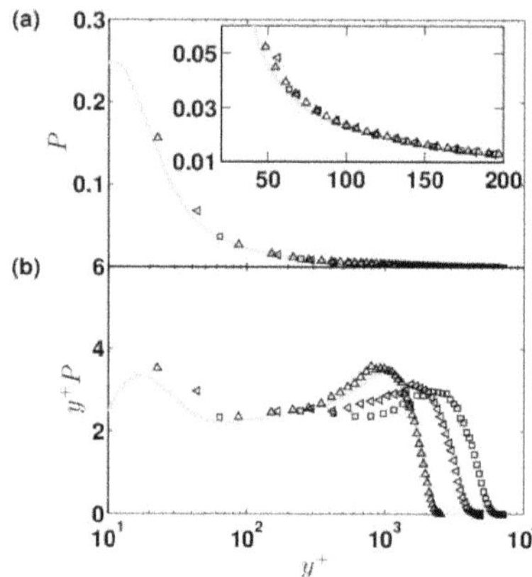


**Figure 3.9:** Profiles of turbulent production (Equation-3.5): —, LES data at $\mathrm{Re}_\theta = 7603$ from Eitel-Amor (2014) and symbols represent the present SPIV data in XY plane; △, $\mathrm{Re}_{\theta,SBL} = 7500$; ◁, $\mathrm{Re}_{\theta,SBL} = 12500$ and □, $\mathrm{Re}_{\theta,SBL} = 18100$; For clarity, one out of every ten data is presented.

Vallikivi et al. (2015a) showed from measurements at high Reynolds number that the skewness profiles of SBL is well collapsed using inner coordinates over the region of $100 < y^+ < 0.15Re_\tau$. They have shown that skewness values are approaching negative values at $y^+ \approx 200$ before finally reach a value of $S \approx -0.1$. Finally, skewness values are more negative in the wake due to intermittency, where they collapse well with the outer coordinates.

Figure-3.10 (a) and (b) shows the skewness $S(u) = \langle u^3 \rangle / \langle u^2 \rangle^{3/2}$ profiles of streamwise velocity plotted with inner and outer scales respectively for SBL cases. Although Figure-3.10(a) exhibit Reynolds number dependence for the outer region when scaled with inner length scales but a fairly good collapse of the same region can be seen in Figure-3.10(b) for all Reynolds numbers when scaled with the outer length scale. In all cases skewness is slightly positive near the wall for $y^+ < 200$ and gradually becoming more negative as the distance from the wall increases. In order to compare the accuracy, LES results at $\mathrm{Re}_\theta = 7603$ from Eitel-Amor (2014) are also plotted as gray continuous line. Present SPIV data agrees well to the similar description of the skewness profiles from Vallikivi et al. (2015a).

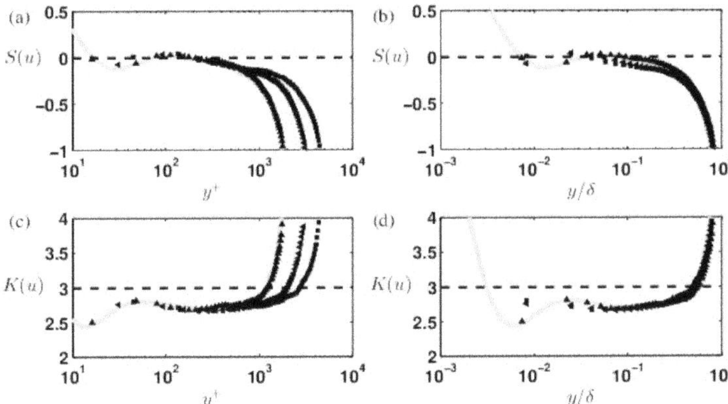


**Figure 3.10:** Skewness and Kurtosis profiles along wall normal direction for SBL cases. —, LES data at $\mathrm{Re}_\theta = 7603$, Filled symbols represent the SPIV data as the same notation used in Figure-3.8. Note that only one out of every five data along wall normal distance is presented for clarity. (a) and (c); inner scaled, (b) and (d); outer one.

The kurtosis $< u^4 > / < u^2 >^2$ profiles for the same data is shown in Figure-3.10(c) with inner scaling and (d) with outer scaling respectively. Inner coordinates shows Reynolds number dependence in the outer region but good convergence is obtained when plotted using outer coordinate. In both cases comparison to the numerical data has an excellent agreement.

Figure-3.11 shows the power spectra of $u'$ from present SPIV measurement for $\mathrm{Re}_\theta = 18094$. Detailed spectral analysis can be found from Foucaut et al. (2004). In order to examine the $k_x^{-1}$ dependence, wall scaled (Vallikivi et al. (2015b)) spectral energy of streamwise fluctuation $\Phi_{uu}$ is plotted against wall scaled spatial wave number ($k_x$) in typical log-log form for $0.045 < y/\delta < 0.1$. A common overlapping region or plateau is observed following -1 slope with fairly good collapse. However, due to the difficulties of converging the spectrum, this -1 slope can not be confirmed as Srinath et al. (2018) suggested that the slope is slightly different from -1 which varies from wall distance. This common region covers almost a decade in the wave number space. Present data is limited to the FoV size and large scales are larger than that of FoV, therefore, large scale energy spectrum was not covered from the present measurements.

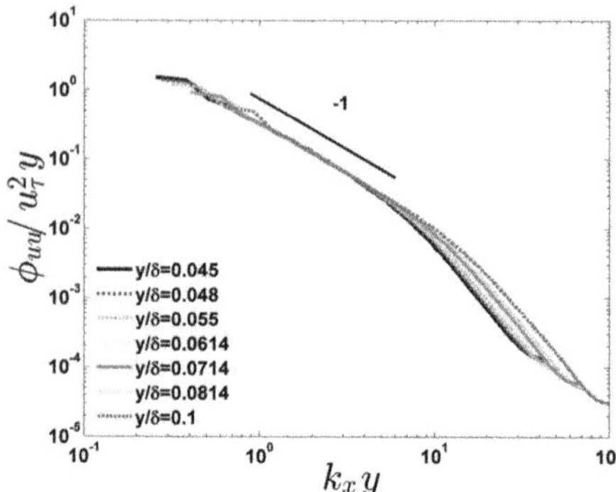


**Figure 3.11:** Wall scaled power spectra of $u'$ at different wall normal location obtained from SPIV data at XY plane for $\mathrm{Re}_\theta = 18094$. The solid black line indicates the $k_x^{-1}$ dependence as guide for the eye.

## 3.3 Micro-blowing results and discussion

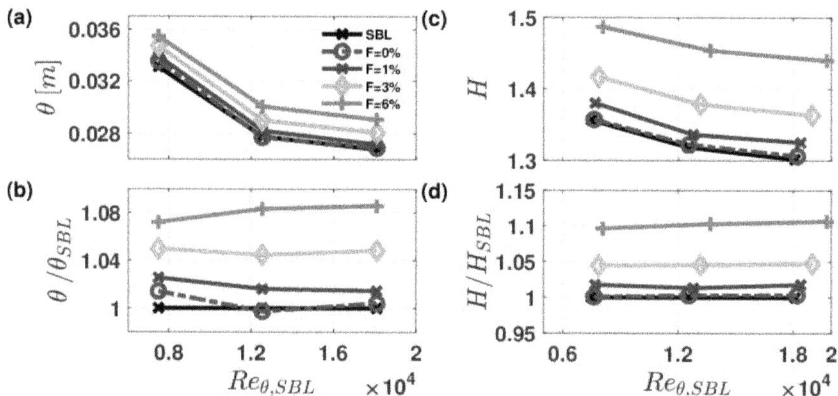


**Figure 3.12:** Variations of momentum thickness ($\theta$) and shape factor ($H$) in comparison to the $\mathrm{Re}_{\theta,SBL}$ for all cases investigated, (a) Momentum thickness (b) Growth of momentum thickness, (c) Shape factor and (d) growth of shape factor in comparison to the SBL cases.

In this section, the changes in boundary layer caused by blowing is introduced. Blowing significantly increases the momentum thickness and the relationship is linear. Although uniform blowing is known for its boundary layer thickening properties, present SPIV data shows that boundary layer growth rate is even stronger at higher Reynolds number.

Figure-3.12 (a) and (b) indicates the increase of momentum thickness the change of momentum loss respectively with blowing ratio (F). Along all the Reynolds number investigated, rate of increase of momentum is almost uniform at a fixed blowing ratio. This indicates that the blowing ratio (F) can be used as an identity parameter in order to compare with the SBL cases as identified by Kametani and Fukagata (2011). Similarly, Figure-3.12 (c) and (d) present the shape factor ($H$) and rate of change with blowing ratio for each case investigated. The shape factor increases with blowing ratios for all cases. As for the momentum thickness, the shape factor increases with blowing rate with same ratio compared to reference case for all Reynolds number investigated (see Figure-3.12d). Combining the observations on $\theta$ and $H$, the increase of blowing ratio determined the increase of boundary layer thickness ($\delta$) independently of the Reynolds number studied. The difference between rate of change of shape factor for SBL and perforated surface without blowing is negligible as inside the uncertainty level. Therefore, it can be inferred that effect of roughness from perforated surface is also negligible at least for the range of Reynolds number measured in the present paper. Kametani and Fukagata (2011) and Kametani et al. (2015) also suggested that blowing increases the boundary layer thickness, momentum thickness and the shape factor. Although, present measurement agrees to the DNS result trend of the mentioned papers, DNS overestimate the growth rate of the stated mean properties. There are two possibilities of such deviation: firstly, growth rate of mean properties reduces with increased blowing or secondly, blowing effect is more prominent at low Reynolds number. The second hypothesis can be privileged as the growth rate of $\theta$ and $H$ with blowing ratio F is nearly linear and independent of momentum Reynolds number.

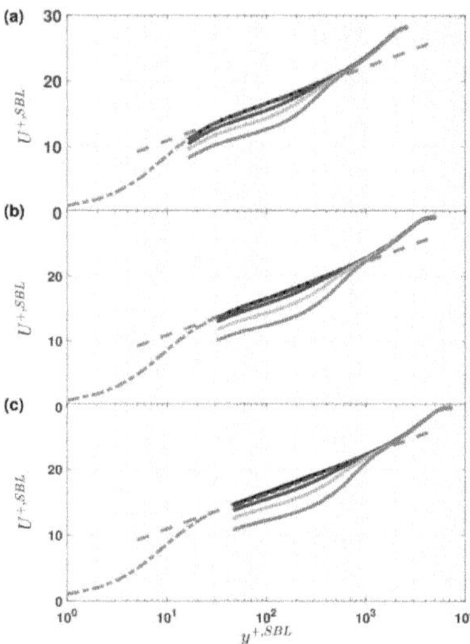


**Figure 3.13:** Mean streamwise velocity profiles for different blowing ratios and momentum Reynolds number studied. In all cases inner variables from the reference SBL cases are used for normalization. Different blowing ratios are indicated with colors as used in Figure-3.12(a). Different Reynolds numbers are depicted in: (a) $\mathrm{Re}_{\theta,SBL} = 7500$, (b) $\mathrm{Re}_{\theta,SBL} = 12500$ and (c) $\mathrm{Re}_{\theta,SBL} = 18100$, Van Driest profile is plotted using 'dashed-dotted' line following Equation-1.30 and logarithmic profile is plotted using 'dashed' line following Equation-3.6 respectively.

Mean streamwise velocity profiles are plotted in Figure-3.13 for the different blowing ratio and momentum Reynolds number studied. As stated earlier in Subsection-3.2.6, wall locations are normalized with the SBL cases which enables the present data to be compared without the biased effect of friction velocity (for the present experiment, friction velocity for the blowing cases were not determined). Most often, measurements of wall shear stress and shear velocity is difficult at high Reynolds number, therefore present scaling helps to ease the process of obtaining the scaling parameters accurately.

Streamwise velocity data is averaged over time and space before plotted as velocity profiles in Figure-3.13 (a), (b) and (c) at $\mathrm{Re}_{\theta,SBL} = 7500$, 12500 and 18100 respectively. Experimental data from Kornilov (2012), DNS and well resolved LES data from Kametani and Fukagata (2011) and Kametani et al. (2015) respectively have already have shown that mean streamwise profile is pushed away from the wall and the effect of blowing is distinct in the outer layer, in particular numerical data exhibit these profiles more prominently where complete profile is influenced eventually. Contrary to the later part of this deduction, present data exhibit that only the inner layer is pushed away from the wall before merging in the wake. This is valid for all blowing ratios at all Reynolds number. Wall roughness effect of the perforated plate without blowing is also found negligible when compared to the SBL cases. This can probably be linked to the large difference in Reynolds number in the present study.

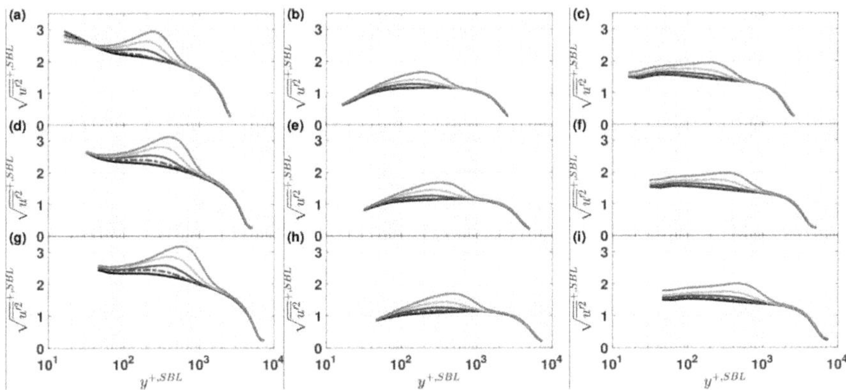


**Figure 3.14:** Root-mean-square of turbulent fluctuations in streamwise, wall-normal and spanwise components. While different blowing ratios are indicated with different colors as used in Figure-3.12(a). (a), (b) and (c) for $\mathrm{Re}_{\theta,SBL}$ = 7500; (d), (e) and (f) for $\mathrm{Re}_{\theta,SBL}$ = 12500; (g), (h) and (i) for $\mathrm{Re}_{\theta,SBL}$ = 18100.

In addition to the mean streamwise profiles, RMS of turbulence fluctuations for all velocity components are presented in Figure-3.14 using inner scales from SBL cases. Here, Figure-3.14 (a), (d) and (g) presents RMS of streamwise velocity fluctuation, (b), (e) and (h) wall normal and (c), (f) and (i) presents spanwise fluctuation respectively. Therefore, turbulence intensities for all three components are enhanced in the outer region with a clear peak between $y^+$ = 200 ~ 500 which grows with blowing ratio. A similar peak appears near this position for high Reynolds number boundary layer links with large scale structures (Hutchins and Marusic (2007b)). It can then be hypothesized that the blowing enhance the large scale structures. Then the effect of blowing diminishes at the beginning of wake region. Present measurement indicates a very interesting phenomena that the near wall region is almost unaffected through blowing while the outer and logarithmic part is strongly altered upto about $y^+$ = 1000.

$$k^{+,SBL} = 0.5\frac{\sqrt{\left((\sqrt{\overline{u'^2}})^2 + (\sqrt{\overline{v'^2}})^2 + (\sqrt{\overline{w'^2}})^2\right)}}{u_\tau} \tag{3.8}$$

Turbulence kinetic energy ($k^{+,SBL}$) is plotted in Figure-3.15 was calculated using Equation-3.8 after normalized with the inner variables obtained from SBL cases. Here, different blowing ratios are indicated with colors as used in Figure-3.12(a). Figure-3.16 (a) plots the positions of outer streamwise component peak values obtained from data used in Figure-3.14. Spatial resolution of the present SPIV system kept the relative error of peak location determination very small, which varied between $y^{+} = \pm 6.5 \sim \pm 18.2$ depending on the reference $\mathrm{Re}_{\theta,SBL}$ from smaller to largest. Each color represent the blowing ratio whereas symbols present particular Reynolds number at SBL. Outer peak of RMS streamwise fluctuation therefore, not only increased in magnitude but also move away from the wall with blowing ratio.

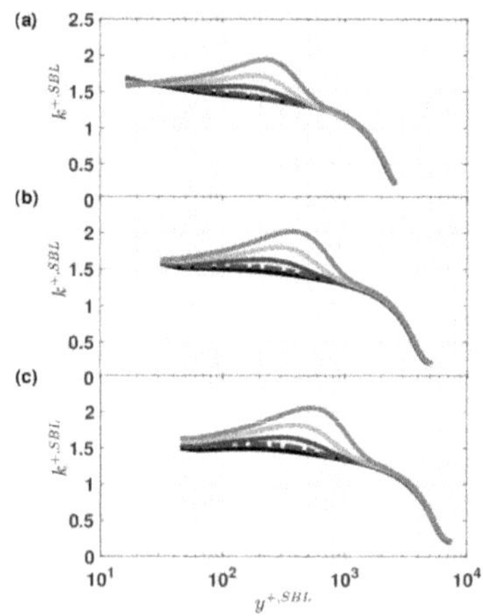


**Figure 3.15:** Logarithmic profiles of $k^{+,SBL}$ along different wall normal locations ($y^{+,SBL}$). Different Reynolds numbers are depicted in: (a) $\mathrm{Re}_{\theta,SBL} = 7500$, (b) $\mathrm{Re}_{\theta,SBL} = 12500$ and (c) $\mathrm{Re}_{\theta,SBL} = 18100$.

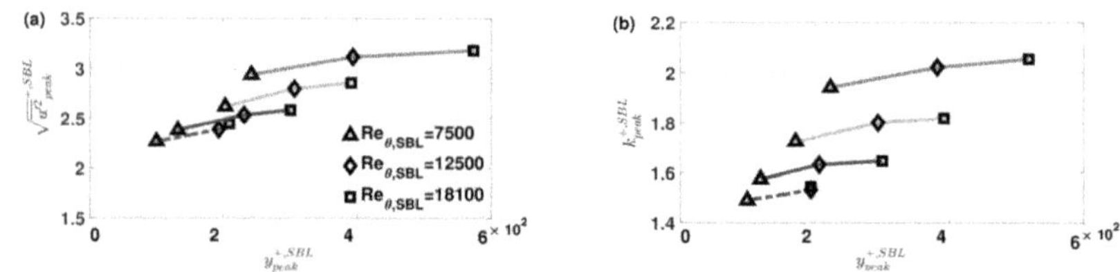


**Figure 3.16:** (a) Outer peak values of $\sqrt{\overline{u'^2}_{peak}}^{+,SBL}$ and their corresponding wall normal position $y^{+,SBL}$ normalized with inner variables obtained from reference SBL cases. Different colors indicate blowing ratio as stated in Figure-3.12 (a). Symbols indicate reference $Re_{\theta,SBL}$, (b) Peak of $k^{+,SBL}_{peak}$ along wall position $y^{+,SBL}$.

Similarly, outer peak of total turbulence kinetic energy $k_{peak}^{+,SBL}$ along corresponding wall location is presented in Figure-3.16 (b). The outer peak of $k_{peak}^{+,SBL}$ follow the same trend as for the outer peak of streamwise component, but with location slightly closer to the wall.

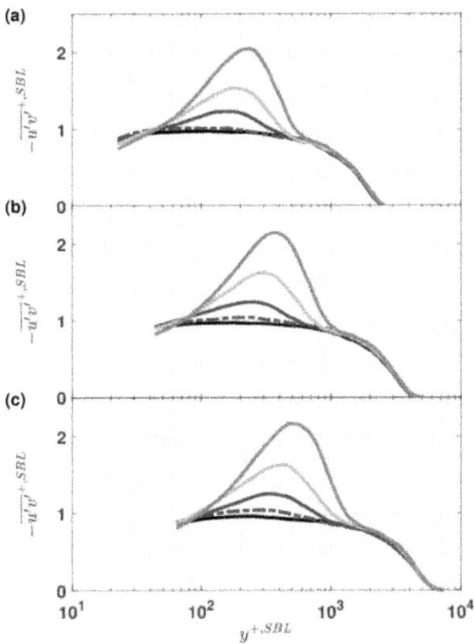


**Figure 3.17:** RSS profiles for the different blowing ratio and Reynolds numbers normalised with inner variables of corresponding reference case, (a) at $\mathrm{Re}_{\theta,SBL}$ = 7500,(b) at $\mathrm{Re}_{\theta,SBL} = 12500$ and (c) at $\mathrm{Re}_{\theta,SBL} = 18100$. Different colors of the profiles indicate different F as indicated in Figure-3.12 (a).

Figure-3.17 shows the profiles of RSS for all investigated case using the similar inner scaling similarly as Figure-3.12 (a). It is observed that a certain part of the outer region is enhanced with a clear outer peak as for all the turbulence intensity components. This peak is getting larger and stronger with increased blowing ratio. The peak value is also shifted away from the wall normal direction as F increases. This enhanced region is also moving with Reynolds number away from the wall. Finally, the profiles merged with their reference SBL data is earlier than predicted by Kametani and Fukagata (2011) who predicted that the full boundary layer is affected. DNS results from this literature described the behaviour of Reynolds shear stress (RSS) that increases with the blowing ratio e.g. in other words, RSS shifted away from the wall and the shifting of the RSS profiles persisted at edge of the boundary layer.

In the behavior of the Viscous Shear Stress ($(\partial U/\partial y)^{+,SBL}$) profiles presented in Figure-3.18, a similar trend in the inner region is observed compared to the results of Kametani and Fukagata (2011) with a decrease of its intensity. However, the trend is opposite in the outer region where the VSS is enhanced. LES data from Eitel-Amor (2014) at $\text{Re}_\theta = 7603$ is also presented in Figure-3.18 (a) as reference for SBL data which shows a perfect agreement. Although difference in the outer region is not much distinctive, therefore a separate plot within each figure is added to highlight the outer region. At all Reynolds number, VSS exhibit a clear peak at highest blowing fraction (F = 6%). On the contrary, VSS profiles are reduced or shifted towards the wall. This phenomena was described as 'counterintuitive'. Although, present measurements are to be found consisted with first part of their findings albeit partially.

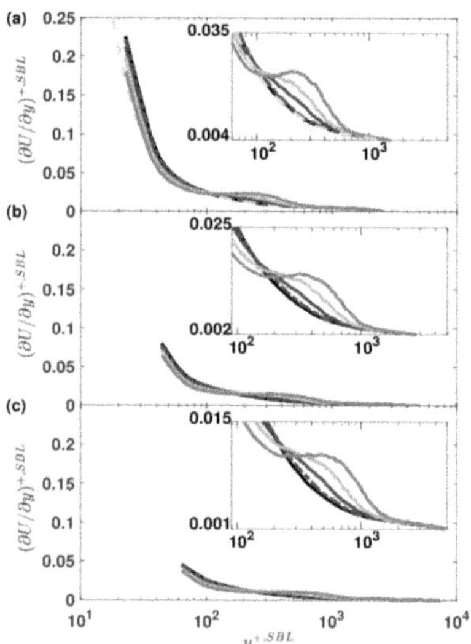


**Figure 3.18:** Profiles of VSS plotted along normalized wall normal distance, different colors indicate blowing ratios in the same manner as Figure-3.12 (a) at $\text{Re}_{\theta,SBL} = 7500$, here, '- - - - -' presents the LES data from Eitel-Amor (2014) at $\text{Re}_\theta = 7603$, (b) at $\text{Re}_{\theta,SBL} = 12500$ and (c) at $\text{Re}_{\theta,SBL} = 18100$.

In order to have a better understanding of the outer peak enhancement by blowing, Figure-3.19 shows $\sqrt{\overline{u'^2}}/\bar{u}$ plotted against $\bar{u}/U_\infty$. This plot can effectively remove the biased effect of wall friction and only the PIV calibration error is the source of error. In such a plot streamwise turbulence intensity scales linearly independent of the Reynolds number for SBL flows. Dotted gray line corresponds to Equation-1.32 (Alfredsson et al. (2011)) where a large part of the log and wake region collapses linearly at $\bar{u}/U_\infty$ between 0.6 and 0.9. A least square fit to the available data from XY plane for SBL conditions did allow us to determine the values of the empirical constants where, a = 0.287 and b = -0.259 which is slightly deviated from Alfredsson et al. (2011) (e.g a=0.286 and b=-0.255). Therefore, SBL data points from the present experiment and reference HWA data from Eitel-Amor (2014) collapse well to the linear relationship from Equation-1.32.

As stated in the description of the figure, markers with different color represent variations in the blowing. Therefore, increased deviation of the data from Equation-1.32 as a result of increased magnitude of blowing is observed. Remarkably data collapses well for each blowing ratio independent of the Reynolds number. Therefore, colored dash-dotted lines present a $5^{th}$ order fitting function to data of blowing. There colors have been chosen as same as the blowing ratio applied to the corresponding data set. In order to have better insight to the deviation from Equation-1.32, outer region is highlighted through the inset figure. Outer layer started deviating from Equation-1.32 at $[\sqrt{\overline{u'^2}}/\bar{u}, \bar{u}/U_\infty] = [0.925, 0.065]$.

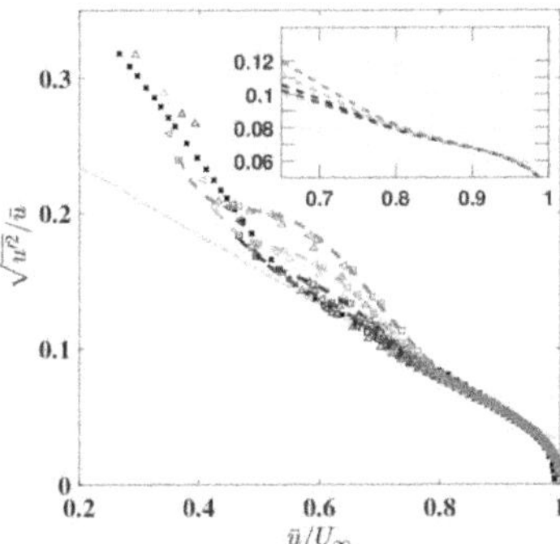


**Figure 3.19:** RMS of streamwise turbulence intensity ($\sqrt{\overline{u'^2}}$) normalized with the local mean streamwise velocity is presented ($\bar{u}$) as a function of $\bar{u}/U_\infty$ after Alfredsson et al. (2011). Here SPIV data from XY plane is presented with hollow markers e.g. △: $\mathrm{Re}_{\theta,SBL} = 7500$; ◁: $\mathrm{Re}_{\theta,SBL} = 12500$ and □: $\mathrm{Re}_{\theta,SBL} = 18100$, '×': HWA data from Eitel-Amor (2014) at $\mathrm{Re}_\theta = 6335$. Different colors indicate variations in blowing ratios where, black symbols represent SBL conditions, blue, violet, green and red represent blowing ratios 0,1,3 and 6% respectively. '-.-' indicate Equation-1.32, '-.-', '-.-, '-.-' and '-.-' indicate a $5^{th}$ order fitting function. Inset figure highlight the region of the intersection of the fitted lines. One out of every 10 data points has been plotted in order to have better clarity.

Figure-3.20 (a), (c) and (e) display the skewness profiles for $\mathrm{Re}_{\theta,SBL} = 7500$, 12500 and 18100 respectively and kurtosis profiles are plotted in Figure-3.20(b), (d) and (f). Blowing induced profiles vary different than that of SBL cases. The skewness varies in the wall normal direction in the log layer from positive values to negative ones for all cases with positive values and negative ones that increase and decrease respectively with blowing ratio. For the maximum value investigated, it varies from 1/2 to -1/2. Finally, all skewness profiles merge to the outer region irrespective of their blowing ratios at certain Reynolds number. On the kurtosis side, similar effects of blowing are observed except the variation is opposite, i.e. Kurtosis first decreases with blowing close to the wall and then increases before merging to each other in the outer part.

Figure-3.21 (a), 3.21 (c) and 3.21 (e) shows the profiles of the main production term in inner scale for all three Reynolds number respectively. Outer region is magnified using a separate plot within the main figure where it can be seen that blowing gradually increases the turbulence production. It is observed that inner and outer region is mostly unaffected while production is maximum within logarithmic region. This is even more prominent in pre-multiplied form used in Figure-3.21 (b), 3.21 (d) and 3.21 (f) which clearly exhibit a peak at all blowing cases. This peak is located within $10^2 < y^{+,SBL} < 10^3$ for all three Reynolds number.

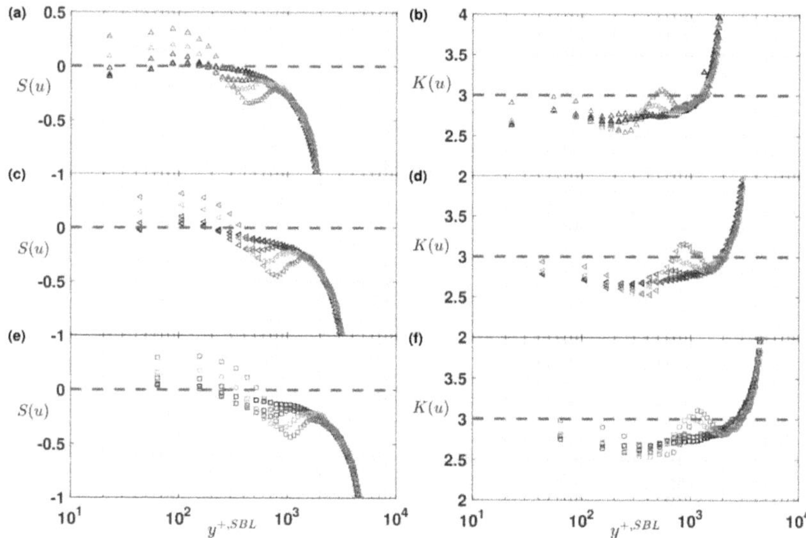


**Figure 3.20:** Profiles of skewness (left column) and kurtosis (right column) of streamwise velocity component (u) along wall normal locations using inner co-ordinates obtained from corresponding SBL condition, '△'; (a) and (b) skewness and kurtosis at $Re_{\theta,SBL} = 7500$ respectively, '◁'; (c) and (d) skewness and kurtosis at $Re_{\theta,SBL} = 12500$ respectively, '□'; (e) and (f) skewness and kurtosis at $Re_{\theta,SBL} = 18100$ respectively. Black symbols represent SBL conditions, blue, violet, green and red represent blowing ratios at 0,1,3 and 6% respectively. One out of every five data points are plotted for clarity.

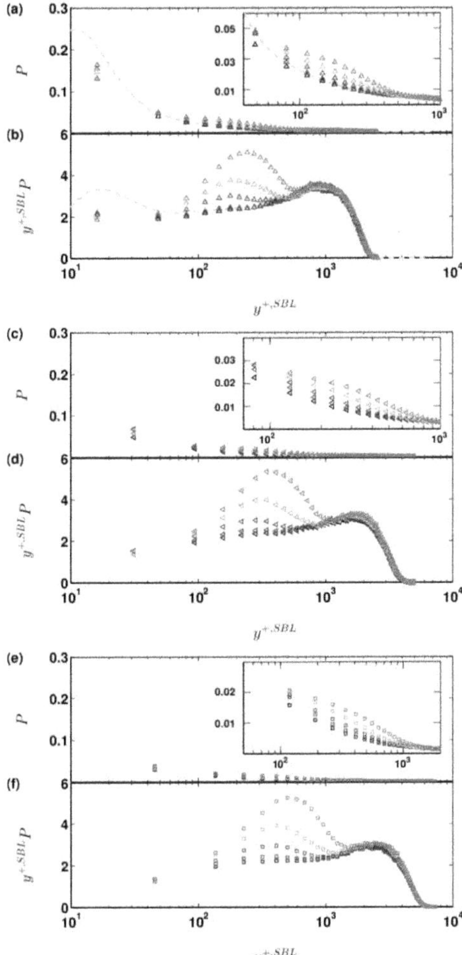


**Figure 3.21:** Profiles of turbulent production term ($P$) and Pre-multiplied production ($y^{+,SBL}P$) along wall locations normalized with SBL inner parameter ($y^{+,SBL}$). '$\triangle$'; (a) and (b) turbulent production term and Pre-multiplied production at $Re_{\theta,SBL}$ = 7500 respectively, where gray 'dashed' line indicate data from Eitel-Amor (2014) at $Re_\theta$ = 7603. '$\triangleleft$'; (c) and (d) turbulent production term and Pre-multiplied production at $Re_{\theta,SBL}$ = 12500 respectively; '$\square$'; (e) and (f) turbulent production term and Pre-multiplied production at $Re_{\theta,SBL}$ = 18100 respectively. For different blowing ratios, color indication is similar to Figure-3.18, One out of every 5 data points have been plotted in order to have better visualization.

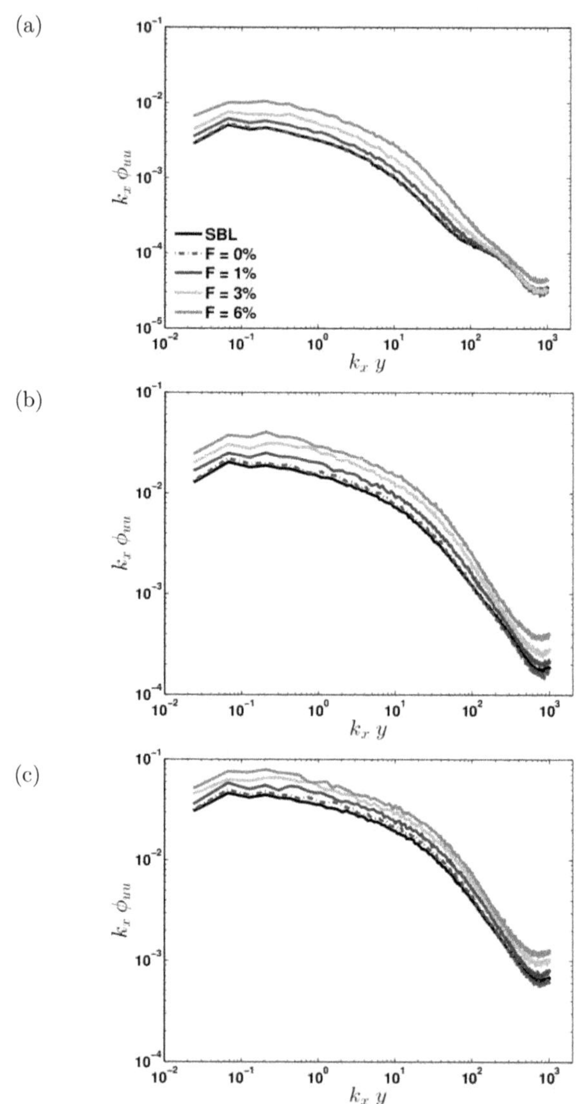


**Figure 3.22:** Pre-multiplied streamwise energy spectrum along wall scaled wave number space. (a) $\mathrm{Re}_{\theta,SBL} = 7500$, (b) $\mathrm{Re}_{\theta,SBL} = 12500$ and (c) $\mathrm{Re}_{\theta,SBL} = 18100$.

Figure-3.22 is plotted in order to look into the pre-multiplied streamwise energy spectrum in the wave number space at $y^{+,SBL} \approx 250$. This is obtained using SPIV measurements following Foucaut et al. (2004). In general, spectral energy increases with increasing blowing ratio. It should be firstly noted that the high wave number range are poluted by noise and the transfert function of PIV (Foucaut et al. (2004)). However, an increase of the pre-multiplied spectrum in this range with blowing ratio for all Reynolds number investigated can be attributed preferentially to an increase of PIV noise probably linked with more out of plane motion due to blowing or by dilution of the seeding concentration of particles by the unseeding air of the perforated plate. Blowing increases the pre-multiplied spectrum at small wave number so the energy contained into large scale structures. Of course, according to the energy cascade, the energy at higher wave number are also enhanced with blowing ratio. Effect of roughness between smooth and perforated plate is found to be small even if the large scale ranges are slightly enhanced.

## 3.4 Conclusion

We have conducted a series of SPIV measurement in a streamwise wall normal plane from a spatially developed turbulent boundary layer manipulated with micro-blowing device. For that a part of the smooth wall condition was replaced with a permeable surface and wall normal blowing was applied for blowing ratio F = 0 ~ 6 %.

In light of the above mentioned results presented we can summarize the results of the present paper in the following categories;

(a) Boundary layer data without blowing was compared with the LES data from Eitel-Amor (2014) as a reference. A very good agreement was found between measured data and the reference LES data upto $4^{th}$ order moment.

(b) The microblowing strongly affect the boundary layer parameter. The momentum thickness and the shape factor are found to increase with blowing ratio. Blowing strongly influence first, second, third and fourth order statistical moments of the streamwise, wall normal and spanwise component of the velocity. Up to fourth order statistical moments for streamwise velocity, up to second order moment for spanwise and wall normal components is presented here for the present study. Blowing strongly affect the near wall region of the mean streamwise profile and enhances the turbulence intensity for all three velocity components with an outer peak which is increasing with blowing and moving away from the wall for all components. Plotting $\sqrt{\overline{u'^2}}/\bar{u}$ versus $\bar{u}/U_\infty$ shows interestingly no Reynolds number dependence at fixed blowing ratio. A fifth order function was fitted on the universal curves obtained at fixed blowing ratio.

(c) Being an active flow control technique, blowing use external energy injected into the flow field, this additional energy effects Reynolds shear stress and Viscous shear stress, eventually lead to an increased production of TKE. Plotting the production term in pre-multiplied form shows that the production is enhanced with blowing in the logarithmic part with a clear peak.

(g) The spectra of steramwise velocity indicates an increase in energy with blowing ratio for low wave number at all three Reynolds number investigated in the present experiment.

Present paper has been written focusing in to the statistical perspective of a turbulent boundary layer under the influence of blowing at different ratios. Measurements from time resolved data with good spatial resolution in spanwise wall normal plane from the same experiment is obtained and currently under analysis. Therefore second part of the present experiment is intended to apply 'frozen turbulence hypothesis' from Taylor and 'attached eddy model' from Townsend and Perry in order to investigate the large scale influence in the log and outer region.

# 4 Coherent motions

In addition to the measurements described in Chapter-3, it was also necessary to identify the flow field immediately above the perforated region. A second set of measurement in SPIV configuration was conducted in order to quantify the effect of uniform blowing adjacent to the blowing assembly. Therefore, the flow was successively measured in YZ plane using SPIV arrangement using a time resolved, high speed SPIV system. A perpendicular in the direction of principle flow e.g spanwise wall normal plane at a streamwise distance of $X_3 = 18.5293\,m$ was designated where the center of the plane was aligned with the wall spanwise center line. Field of view was set immediately over the perforated surface at 25% downstream from the beginning.

FoV in this plane was $0.09 \times 0.061 m^2$ in spanwise and wall normal direction respectively as indicated in Figure-4.2(a). 832 × 768 pixel$^2$ camera resolution was applied in the respective directions as stated earlier. Description of this set-up will be discussed in following subsections.

## 4.1 Experimental setup

LMFL boundary layer wind tunnel was used as described in Chapter-3. Therefore, no additional description of the wind tunnel is added in this chapter.

### 4.1.1 Wind tunnel instrumentation

Upstream blowing with uniform velocity in a flat plate TBL was established with the same perforated plate as explained in Chapter-3. Spatially developed turbulent boundary layer over a flat plate TBL was established with a perforated region where 4514 holes with uniform diameter of 3.6 mm were constructed with staggered arrangement. Figure-4.1(a) shows the location of the perforated region. For incompressible TBL with larger length scales, necessary modifications of the blowing assembly was done compared to the design data from Hasanuzzaman et al. (2016). In order to provide wall normal blowing, a solid wind tunnel wall was replaced with a perforated (blowing surface) one and the air supply was provided as shown in Figure-4.1(b). Streamwise length of the blowing surface was at wind tunnel characteristics length $X = 18.425 \sim 18.845$, keeping the width center equidistant from both side walls of wind tunnel. Blowing rate is expressed as blowing ratio (BR) was applied at a very low velocity (0, 1, 3 and 6%) for each Reynolds number being measured.

The flow was seeded with tracer particles by a Hazebase Base Classic fog machine located in the wind tunnel diffuser section. Particles are globally provided by a repeating evaporation and condensation process using a solution of poly-ethylene-glycol and water. The mean particle diameter is approximately 1 $\mu m$ and have a lifetime of around 10 minutes circulated within the closed circuit wind tunnel.

(b) 

(a) 

**Figure 4.1:** (a) Photograph showing the location of the MBT surface where the white arrow indicate the distance from the leading edge, here, direction of the flow is from right to left from the readers perspective and (b) Photographs of the bottom side the MBT surface showing air supply lines.

### 4.1.2 SPIV in YZ plane

2 × Phantom Miro 340 cameras each equipped with a Nikon Nikkor 200 mm were placed in stereo arrangement and forward scattering mode from the same side of the tunnel measurement section as shown in Figure-4.2(b). Cameras were placed with an angular stereo configuration. Although,Prasad (2000) recommended an off axis stereo angle of $45^{\circ}$ in order to get a maximum accuracy of the out of plane component relative to the in plane components. However, for YZ plane, the stereo angle between 2 cameras were set to a nominal value of $40^{\circ}$. CMOS resolution of both the cameras were reduced to 768 × 832 pixel$^2$ along wall normal and spanwise axis in order to realize high rate of image acquisition at $f_{acq} = 2$ kHz. Therefore, each pixel in the image plane corresponds to a distance in object space for 76.35 $\mu$m and 126 $\mu$m in wall normal and spanwise direction respectively.

From Figure-4.2, the blue line indicate the spanwise length of the FoV in top view projection. The plane is $X_3 = 18.5293\ m$ from the plate leading edge and placed perpendicularly over the $14^{th}$ hole row. Internal RAM for each camera was saving the images at a 8-bit of dynamical resolution of the CMOS sensor array, however, after the image transfer acquired images were decompressed having 16-bit of dynamical resolution without losing any data. Although, full resolution of the camera could not be used in order to optimize the image acquisition frequency ($f_{acq}$). Therefore, at $f_{acq} = 2$kHz, camera resolution was reduced to 768 × 832 pixel$^2$ where the larger side of the camera sensor was employed to image the spanwise (z-axis) extent of the flow field.

A Quantronix Darwin Duo laser with 20 mJ/pulse was used to generate laser to illuminate the designated plane as a light sheet passing through a set of optical lenses and 2 mirrors placed in 45°. The laser light sheet was incident from the transparent side walls of the wind tunnel test section as indicated in Figure-4.2(a), edge of the light sheet was immediately adjacent to the wall so as to avoid reflection from the wall. Therefore, laser energy were adjusted in order to achieve sufficient amount of energy being imparted to the particles present in the FoV. Similarly, laser optics were also adjusted in order to obtain an uniform laser sheet thickness of 1 mm.

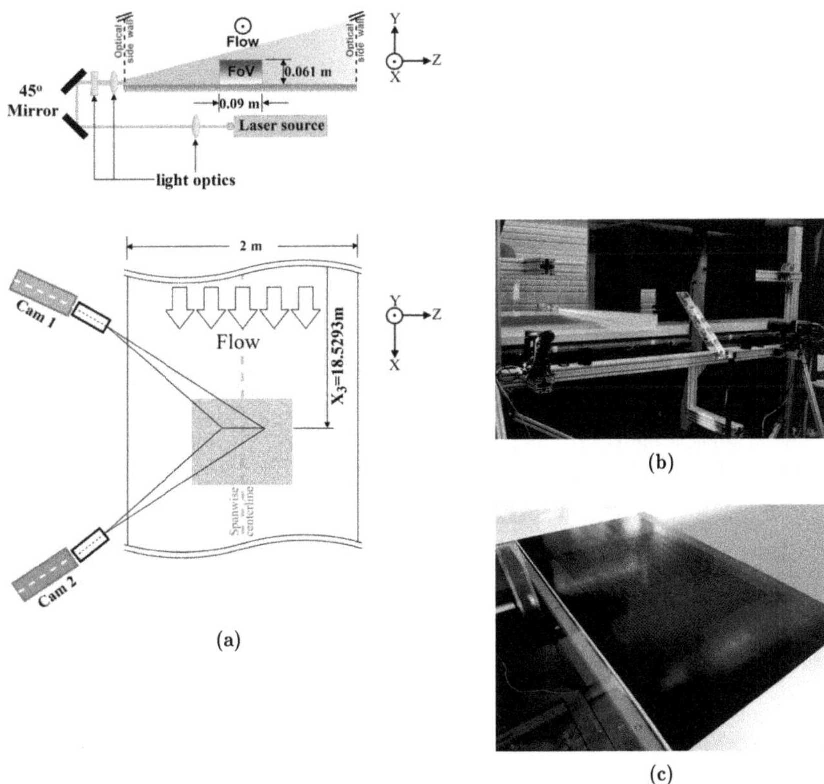


**Figure 4.2:** (a) Front view schematic of the YZ plane (top), Top view schematic of the YZ plane (bottom); (b) Photograph of the test section in streamwise wall normal orientation, SPIV arrangement for YZ plane, the flow is coming from left to right relative to the reader and (c) Photograph of the perforated plate attached to the wall.

| | |
|---|---|
| SBL Reynolds number ($Re_{\theta,SBL}$) | 7495 |
| | 12542 |
| | 18094 |
| Plane | YZ |
| Stereoscopic angle | 45° |
| Focal Length (mm) | 200 |
| Laser sheet thickness (mm) | 1 |
| Magnification | 2 |
| Resolution ($\mu m\, px^{-1}$) | 100.351 |
| Lens aperture ($f\#$) | 11 |
| Camera distance from FOV (m) | 2 |
| Field of View (m) | [0.061,0.09]($S_{y,2}$, $S_{z,2}$) |
| CMOS array (px × px) | 768 × 832 (y,z) |
| Image acquisition frequency (Hz) | 2000 |

**Table 4.1:** PIV recording parameters

Figure-4.2(b) shows the photograph of the test section from the camera side which actually, is the isometric photographic representation of the Figure-4.2(a) (top). This displays the laser coming from the side glass of the test section being tangential over the wall in YZ orientation along with the camera positioning. Figure-4.2(c) represents the photograph of the perforated region indicated with red color in Figure-4.2(a)(bottom).

Used SPIV setup was capable of high frequency measurement. In order to obtain time resolved velocity data of the designated plane, image acquisition at 2 kHz was done for a period of 3.2 seconds per run. As a consequence, 6400 image samples were obtained for each run. For YZ configuration, mean flow is normal to the plane of measurement which may lead to large out of plane motion. Therefore, two laser sheets were separated by a distance of 500 $\mu m$ (mean out of plane displacement) in order to avoid large out of plane motion, which essentially improves the Signal to Noise Ratio (SNR) and simultaneously, maintain a high order of cross correlation (Foucaut et al. (2014)). Separation time between the two cavities ($\Delta t$) was chosen for a maximum in-plane displacement of around 10 pixels for each Reynolds number being measured. Subsequently 4 independent runs per cases of blowing at each Reynolds number were acquired which consists of a total of 48 runs. But only the selected results from the measured data will be presented in this paper. Table-4.1 summarizes the salient aspects of the SPIV systems used for the present experiment.

### 4.1.3 Evaluation of PIV

A nomenclature stated in Table-4.2 indicate in a sequence the reference plane of measurement, free stream velocity, surface condition and rate of blowing fraction applied. Where, first 2 letters in capital 'YZ' indicate reference plane of measurement from spanwise-wall normal plane, third letter 'U' followed by a fourth or fifth digit indicate free stream velocity and the fourth letter 'S' or 'F' indicate surface condition which is 'SBL' or blowing respectively. Finally, the last digit indicate the rate blowing fraction applied.

| $Re_{\theta,SBL}$ | $U_\infty$ [m/s] | BR ($\%U_\infty$) | YZ plane |
|---|---|---|---|
| 7495 | 3.4 | smooth | - |
| | | 0 | YZU3F0 |
| | | 1 | YZU3F1 |
| | | 3 | YZU3F3 |
| | | 6 | YZU3F6 |
| 12542 | 6.8 | smooth | - |
| | | 0 | YZU6F0 |
| | | 1 | YZU3F1 |
| | | 3 | YZU3F3 |
| | | 6 | YZU3F6 |
| 18094 | 10.2 | smooth | - |
| | | 0 | YZU10F0 |
| | | 1 | YZU10F1 |
| | | 3 | YZU3F3 |
| | | 6 | YZU3F6 |

**Table 4.2:** PIV recording nomenclature

### 4.1.4 Image evaluation

Images were processed using an in house LMFL modified version of MatPIV code. Image deformation due to stereoscopic aberration was adjusted using the Soloff back projection/re-construction of three dimensional warping technique (Soloff et al. (1997)). This is the same 3D warping technique described in Coudert and Schon (2001) and implemented in the calibration process described as "Self-calibration" applied in Wieneke (2005). Laser light sheet misalignment correction was done by cross-correlating the two PIV mapped images, which gives us the opportunity to correct the error on the 3D vector origin.

PIV evaluation was first performed on the image space of camera. For the stereo system, 2 images taken by each camera was then interpolated using 2 frame cross correlation in order to identify displacement of particles. Later, overlapping of the generated mesh for each stereo systems were created. Common grid points from the calibration target is then reconstructed using 2D2C component. Number of mesh points for the stereo system had a 114 points along x-axis and 74 points along y-axis with an increment of 0.0008 m along both the axis (corresponding to 7.435 pixel along x-axis or 9.682 pixel along y-axis), where, first mesh point along y-axis was determined at 0.0012 m or 14.5

| Plane (1-2) | $Re_{\theta,SBL}$ | $\delta^+$ $Re_\tau$ | $u_\tau$ (m/s) | FoV $S_1$, $S_2$ | Interrogation Window [$L_{IW}$] | Mesh step [$\Delta_i/l^+_{SBL}$] | No. of records |
|---|---|---|---|---|---|---|---|
| yz | 7495 | 2186 | 0.1211 | $0.23\delta$, $0.34\delta$ | $14.8^+ \times 18.4^+$ | $6.5^+$ | 3600 × 4 |
| yz | 12542 | 3780 | 0.2345 | $0.25\delta$, $0.37\delta$ | $28.7^+ \times 35.6^+$ | $12.5^+$ | 3600 × 4 |
| yz | 18094 | 5482 | 0.3455 | $0.26\delta$, $0.38\delta$ | $42.3^+ \times 52.4^+$ | $18.5^+$ | 3600 × 4 |

**Table 4.3:** Reference case characteristics for SPIV processing

pixels from the wall.

The MatPIV code stated earlier of this section is a Fast Fourier Transform based PIV analysis algorithm for MATLAB using multiple grid and multiple pass. Particle images were first evaluated in the image space of each camera. Later, images are divided into discrete windows in the pixel array by the order of integer segments which is also known as 'Interrogation Windows (IW)' or 'Sampling Window (SW)'. Thereafter, normalized cross correlation function between 2 IWs of a double field image is calculated following Soria et al. (1999). Eventually, particle displacement in the pixel array of the particular IW is determined by least-square-fitting of Gaussian distributions of extremum from the cross-correlation values. Therefore, accuracy of the peak detection depends on the quality of acquired images and the amount of particle displacement. For the presented result, magnification of the lenses and maximum displacement of the particles were selected as such that optimized light scattering from the particles were obtained (Willert and Gharib (1991)). In addition, mapping function (Soloff et al. (1997)) of particle displacement from image plane to object plane is done with the 2D calibration based reconstruction.

Images evaluation of the data sets were performed with standard multiple grid algorithm and discrete window offset (Westerweel et al. (1997) and Soria et al. (1999)). In the sequence of image processing, 4 successive passes were used with an initial window size of 64 × 64 pixel$^2$ to a final window size of 18 × 24 pixel$^2$. In between the initial and the final pass, 2 successive passes with a IW size of 32 × 32 pixel$^2$ cross correlation analysis were also performed. 60% overlapping of the IW of the successive passes were used to increase SNR and displacement estimation which eventually improve the performance of FFT based cross correlation function (Westerweel et al. (1997)). Additional image deformation was applied in order to compensate the IW stretching before the final pass following the iterative adaptive resolution scheme (Scarano (2002)).

Final IW size determines the spatial resolution of the present measurement when expressed in wall units. Final IW varied between $14.8^+ \sim 52.4^+$ depending on the Reynolds number of the flow. On the other hand, IW from YZ plane posses different length along y and z-axis. Table-4.3 illustrates the different post processing parameters for the reference cases. Length of IW ($L_{IW}$) is given in the order of the axis from plane description.

Here, IW is stretched in the spanwise direction following the $IW_{ar}$.

## 4.2 Validation

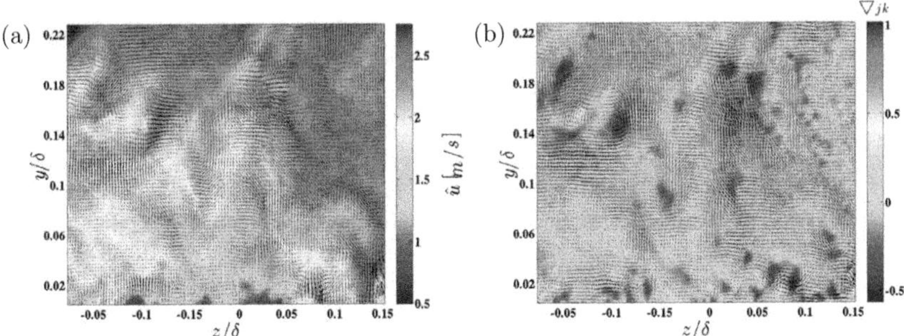


**Figure 4.3:** (a) Instantaneous velocity contour plot of the YZ plane in $Re_{\theta,SBL}$ = 7495; (b) Instantaneous vorticity contour plot of the same snapshot.

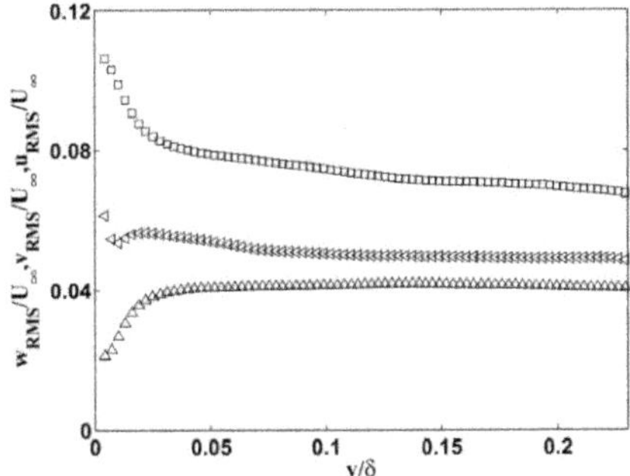


**Figure 4.4:** At $Re_{\theta,SBL}$ = 7495 RMS values of fluctuations normalized with $U_\infty$ along different wall normal height scaled with $\delta$ (measurements from YZ plane), □ : $u_{RMS}/U_\infty$, ◁ : $v_{RMS}/U_\infty$, △ : $w_{RMS}/U_\infty$.

Figure-4.3(a) shows the contour plots of the instantaneous streamwise velocity normalized with the free stream velocity of a snapshot. Arrows indicate the velocity vectors obtained from the spanwise and wall-normal velocity components. Vertical and horizontal axis represent the outer scaled wall normal and spanwise distances in Cartesian co-ordinate (FoV). Figure-4.3(b) presents the same snapshot where color scheme applied to show the spanwise and wall-normal vorticity with arrows indicating streamwise wall-normal velocity. In both of the figures, 0 indicate the spanwise center of the test section. These figures visually represent that the spatial resolution of the SPIV system was excellent in order to resolve the small scale vortices present in the flow.

High speed SPIV acquisition was realized in order to obtain time correlated data with sufficiently high frequency ($f_{acq}$ = 2 kHz). For each Reynolds numbers and blowing ratios, 4 runs were acquired. RMS of the different velocity components scaled with the free stream velocity along ascending wall normal height as a fraction to $\delta$ is presented with Figure-4.4 at $Re_{\theta,,SBL}$ = 7495. We can observe that $\sqrt{\bar{v}'}$ is maximum at $y$ = 0.0044 m or $y/\delta$ = 0.016, which is the location where natural peak value is found at the measured Reynolds number. Blowing as an active method, is expected to add energy to the wall normal component. The addition of energy is dependent on the Blowing Ratio. At the same time, wall normal component is expected to curtail the magnitude of the streamwise velocity changing the mean gradient ($du^+/dy^+$) of $\bar{u}$ at the near wall region ($y^+ \leq 5$).

## 4.3 Results

Figure-4.5(a)-(d) and 4.5(a)-(d) indicate the contour plots of streamwise and wall-normal fluctuation respectively. Here, streamwise fluctuations are normalized with $U_\infty$ and cartesian co-ordinates along z and x axis were normalized with $\delta$. In order to obtain spatial distribution of the streamwise velocity fluctuations presented in Figure-4.5 (a)-(d), Taylor's frozen turbulence hypothesis has been employed to infer the spatial velocity field from the temporal SPIV data. Interested readers are advised following the space time conversion procedure from Monty et al. (2007).

Figure-4.5 presents contour plots of streamwise velocity fluctuations measured with the high temporal resolution SPIV in YZ plane at $Re_{\theta,SBL}$ = 7495. The streamwise velocity has been scaled with the free stream velocity, $U_\infty$. (a) BR = 0; (b) BR = 1%; (c) BR = 3% and (d) BR = 6%. Similarly, Figure-4.6, exhibit the wall normal fluctuation. Taylor's frozen turbulence hypothesis was inferred in order to obtain the contour plots at $y/\delta$ = $1.62e^{-06}$ ($y^{+,SBL} \approx 36$).

Recent study from Ganapathisubramani et al. (2007) advocated that Taylor's hypothesis should not significantly deviate the large scale structures as can be observed from Figure-4.5 and 4.6. The blue low-speed regions surrounded by red high-speed regions are the signature of the Coherent motions in TBL. Corresponding wall normal height is within the logarithmic and the beginning of the wake. In some cases, spanwise length of such motions exceed the length of FoV. Here, regions indicated by blue, flanked by red is the signature of high velocity turbulent spots, which forms a larger packet of smaller spots as the blowing increases. Gradually, their occurrence grows as blowing rate increases, therefore, large regions with stronger energy are observed in Figure-4.5 (d). Similarly, wall normal fluctuation exhibit more low speed 'spots' which was quite unexpected. Wall normal velocity applied at wall is more likely to contribute to the streamwise components rather than adding to the vertical component of velocity. These results are pretty much in line to the results from Kametani et al. (2015) and Stroh et al. (2016)

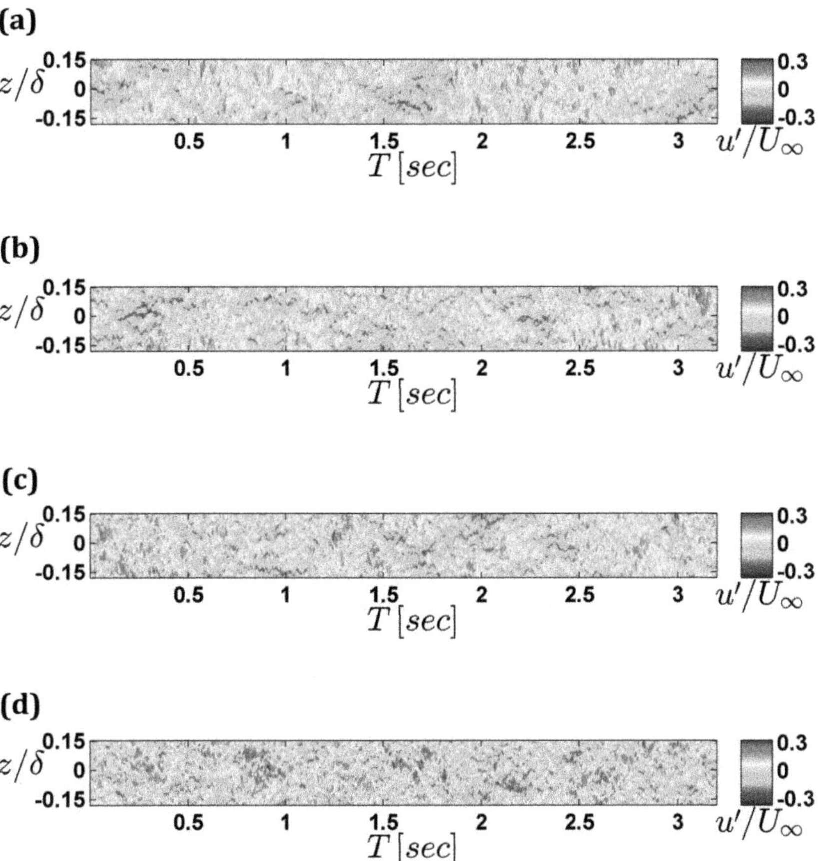


**Figure 4.5:** Contour plots of streamwise velocity fluctuations normalized with $U_\infty$ over perforated surface at $Re_{\theta,SBL}$ = 7495, distance from the wall $y/\delta$ = $1.62e-06$ ($y^{+,SBL} \approx 36$). (a), (b), (c) and (d) presents different blowing ratios F = 0, 1, 3 and 6 % respectively. Flow is coming from left to right with the readers reference point.

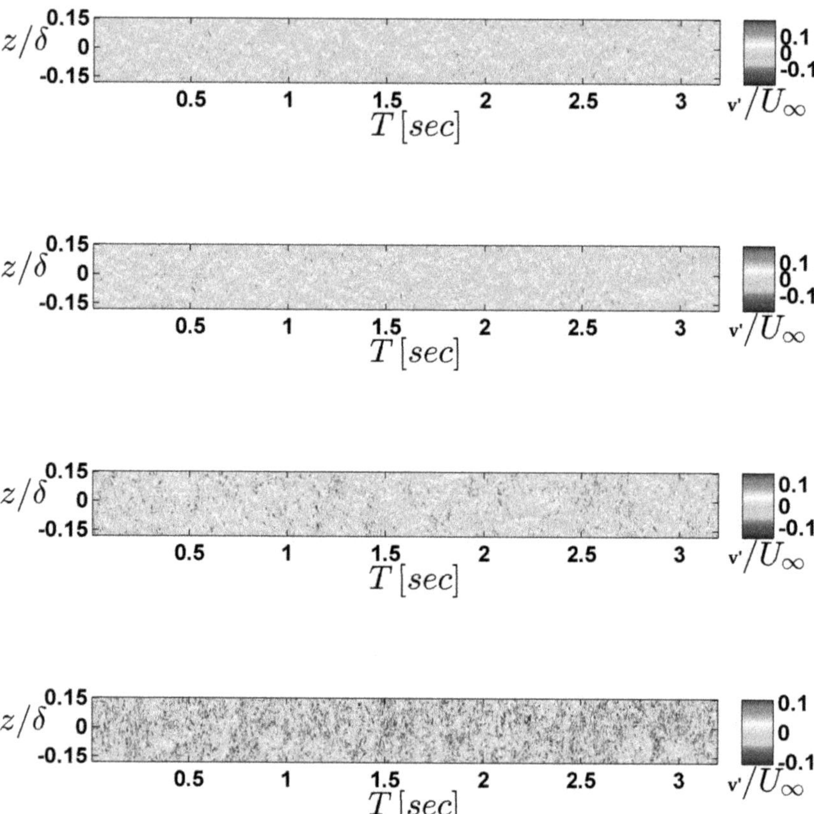


**Figure 4.6:** Contour plots of wall-normal velocity fluctuations normalized with $U_\infty$ over perforated surface at $Re_{\theta,SBL}$ = 7495, distance from the wall $y/\delta$ = 1.62$e-$06 ($y^{+,SBL} \approx 36$). (a), (b), (c) and (d) presents different blowing ratios F = 0, 1, 3 and 6 % respectively. Flow is coming from left to right with the readers reference point.

# 5 Concluding remarks

A series of turbulent boundary layer profiles over flat plate geometry was measured at two different wind tunnels with no pressure gradients. A wide range of Reynolds number was investigated where moderate Reynolds number range was investigated at the BTU wind tunnel and subsequently, LMFL boundary layer wind tunnel used for the high Reynolds number ranges.

## 5.1 Moderate Reynolds number experiment

Chapter-2 presents the measurements at moderate Reynolds number, where measurements were carried out using LDA technique. Profiles of streamwise and wall-normal velocity components were obtained at different Reynolds number ($Re_{\theta,SBL}$ = 1100 ~ 3670). Uniform blowing was applied using a finite perforated wall where measurement location was directly above the perforated plate. Three different blowing velocity was applied and BR was varied between 0.17 ~ 1.52. Estimated error on the average to the reference smooth wall measurements were within less than 0.5 % of $U_\infty$. Wall shear was determined using direct measurements of the near wall data using LDA. Therefore, profiles of velocity components such as mean, Root-Mean-Square, skewness and kurtosis were presented using both the viscous length scales and the outer scale parameters. Mean profiles show significant deviation in shape in the region $y/\delta$ = 0.005 ~ 0.55, where streamwise velocity is suppressed and simultaneously, wall normal velocity is enhanced. Both the effects depend on the magnitude of blowing applied.

The profiles of mean velocity scaled with their corresponding wall shear and kinematic viscosity (viscous length scale) were plotted in semi-logarithmic plots. It was observed that the profiles show overlapping within the viscous sub-layer region ($y^+ \leq 5$) for different BRs. However, profiles for blowing induced TBL start deviating in the buffer region shows clear accent. The strength of the accent depends on the BR and the behaviour is proportional. This suggest that a *logarithmic law* cannot be used to effectively describe the mean-velocity profile of boundary layers under the influence of uniform blowing. *Velocity defect-law* presentation of the data also shows gradual deviation of the profiles depending on the BR. Therefore, both of the traditional way of presenting mean streamwise velocity profiles can not be used for the blowing induced TBL.

However, mean streamwise velocity was plotted independent of their shear velocity following the *diagnostic plot*. Although, we cant obtain an independent overlapping when profiles of different blowing ratios were plotted at the same reference Reynolds number.

RMS of the streamwise fluctuation was plotted in the semi-logarithmic axis using inner scaling parameters. Inner peak was found at the wall distance suggested by the literature. However, blowing enhances the so-called outer peak and imitates the outer peak found at the higher Reynolds numbers. The plateau of the outer peak accent with the magnitude of the BR. Effect of blowing on the RMS of the streamwise fluctuation is prominent at the logarithmic region.

Inner scaled RMS of the wall-normal fluctuation is also enhanced in the logarithmic region. The plateau is roughly uniform along wall distance within the logarithmic region.

Although, near wall measurements for blowing turbulent boundary layer is a daunting task and subjected to high order of uncertainty, however, within this experiment we can obtain direct wall shear stress measurement with the help of a non-intrusive technique with high accuracy. We have ascertained the selection of near wall data points based on the correlation coefficient, which subsequently detects the linear region due to viscosity in the sub-layer region. Friction co-efficient was calculated from the wall shear stress and blowing was found to be reducing skin friction. However, rate of skin friction reduction depends on the BR. Slope of the mean streamwise velocity ($d\bar{u}/dy$) reduces as the blowing increases. Flow control using uniform blowing is an active means of flow control technology. Therefore, in order to relate the drag reduction to the energy input, effect of blowing was analysed with the help of performance indicators. Drag reduction rate is not proportional to the BR. However, the reduction rate reaches an optimized value which corresponds to a specific BR. However, this has a Reynolds number dependence and therefore, requires further study. Net energy saving rate was also presented along the net gain based on the local friction co-efficient data. Presented data is in close agreement with the numerical data from Kametani et al. (2015). Although, limited number of blowing ratios were investigated. However, in order to reach a definite conclusion, a wide range of blowing ratios are advised to investigate.

## 5.2 Spatially developed TBL at high Reynolds number

High resolution PIV measurements were conducted for a spatially developed large Reynolds number TBL. Detailed description of the experiment is already published as Hasanuzzaman et al. (2020) and has been added to this thesis as Chapter-3. Three different Reynolds number were investigated at fixed streamwise distance. High fidelity SPIV data has been presented with a good accuracy for the statistics upto fourth order moments. BR investigated were 0, 1, 3 and 6% of $U_\infty$ respectively. This experiment is unique because of the large spatial boundary layer thickness of the turbulent boundary layer. Therefore, large boundary layer thickness allowed us to measure all three components of velocity. Reference Standard Boundary Layer data was compared together with the data from Örlü and Schlatter (2013) and Eitel-Amor (2014) and was found in excellent agreement. A detailed conclusion is also presented at the end of the literature, therefore require no further information.

## 5.3 Coherent motions

Chapter-4 presents the second experiment conducted at the LMFL wind tunnel. Accuracy of the results are in good agreement with literature values. Although, accuracy of the data is strongly depending on the accuracy of the images. Therefore, PIV error was not more then 0.1 pixels.

Contour plots of streamwise velocity based on Taylor's frozen hypothesis exhibit small packets of high velocity region which increases with blowing. On the contrary, low velocity packets are in dominant number for wall normal velocity. Therefore, blowing air affects streamwise velocity in adding momentum whereas it prevents the wall normal component (Figure-4.5 and 4.6). Results from this experiment help us to explain the mechanism of the blowing and proves the presented hypothesis in Chapter-1 partially.

## 5.4 Outlook

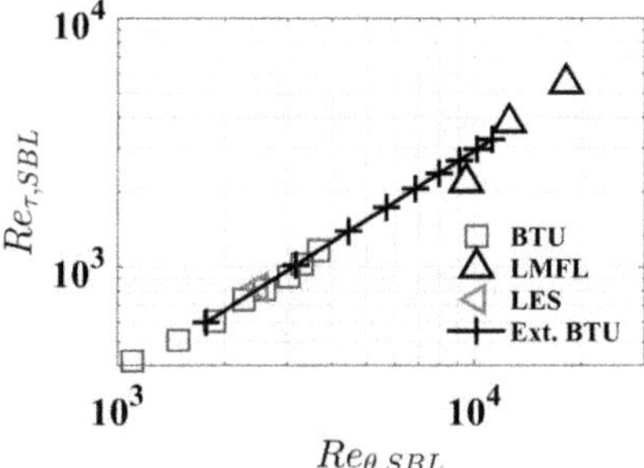


**Figure 5.1:** Proposed extension of the experimental ranges at BTU.

Figure-5.1 is the re-plotting of Figure-1.22 from Chapter-1(a). Present thesis has covered an wide range of Reynolds number, however, further experiments are required in order to investigate the Reynolds number ranges filling the gap between the BTU and LMFL wind tunnel facility.

# Bibliography


Albrecht, H.-E., Damaschke, N., Borys, M., Tropea, C. (2003). *Laser Doppler and Phase Doppler Measurement Techniques, Springer verlag*, 738.

Atzori, M.; Vinuesa, R., Fahland, G., Stroh, A., Gatti, D., Frohnapfel, B. and Schlatter, P. *Aerodynamic Effects of Uniform Blowing and Suction on a NACA4412 Airfoil., Flow Turb. Comb., doi:10.1007/s10494-020-00135-z*, 105:735 – 759.

Airbus Press Release. *Airbus "BLADE" laminar flow wing demonstrator makes first flight, www.airbus.com*, https://www.airbus.com/newsroom/press-releases/en/2017/09/airbus_-_blade_-laminar-flow-wing-demonstrator-makes-first-fligh.html, (2020, January 1).

Anders, J., (1989). *LEBU drag reduction in high Reynolds number boundary layers. 2nd Shear Flow Conference, March 13-16*, Tempe, AZ, U.S.A., 1 – 10.

Abbassi, M. R.,Baars, W. J., Hutchins N. and Marusic I., (2017). *Skin-friction drag reduction in a high-Reynolds-number turbulent boundary layer via real-time control of large-scale structures. Int. Jour. Heat and Fluid Flow, 67*, 30 — 41.

Alfredsson, P. H. and Örlü, R., (2010). *The diagnostic plot–a litmus test for wall bounded turbulence data. Eur. J. Mech. B/Fluids, 29*, 403 -– 406.

Alfredsson, P. H., Segalini, A. and Örlü, R., (2011). *A new scaling for the streamwise turbulence intensity in wall-bounded turbulent flows and what it tells us about the "outer" peak. Phys. Fluids, 23*, 041702.1 -– 4.

Ahn, S. and Fessler J., A., (2003). *Standard errors of mean. Variance and standard deviation estimators.* http://web.eecs.umich.edu/~fessler/papers/files/tr/stderr.pdf.

Adrian, R., J., (2007). *Hairpin vortex organization in wall turbulence, Phys. Fluids*, Vol-19, pp. 1 – 16.

Adrian, R., J., Christensen, K., T. and Liu, Z-C., (2000). *Analysis and interpretation of instantaneous velocity fields, Exp. Fluids*, Vol-29, pp. 275 – 290.

Adrian, R., J., Meinhart, C., D. and Tomkins, C., (2000). *Vortex organization in the outer region of the turbulent boundary layer, J. Fluid Mech.*, Vol-422, pp. 1 – 54.

Blackwelder R., F. (2007). *Some ideas on the control of near-wall eddies. AIAA, Paper No. 89-1009.*

Burden, H., W., (1970). *The effect of wall porosity on the stability of parallel flows over compliant boundaries, Naval Ship Research and Development Center*, Tech. Rep. 3330.

Beck, N., Landa, T., Seitz, A., Boermans, L., Liu, Y. and Radespiel, R. *Drag reduction by laminar flow control*, *Energies*, 11 (252), 1 – 28, 2018.

Banister, D., Anderton, K., Bonilla, D., Givoni, M. and Schwanen, T. (2011). *Transportation and the environment.Annu. Rev. Environ. Resour.*, *36* , 247 -– 270.

Benedict, L., H. and Gould, R. D. (1996). *Towards better uncertainty estimates for turbulence statistics.Exp. Fluids, 22(2)* , 129 – 136.

Bushnell, D. M. and Tuttle, M. H., (1979). *Survey and bibliography on attainment of laminar flow control in air using pressure gradient and suction. NASA, RP – 1035.*

Bushnell, D. M., (1983). *Turbulent Drag Reduction for External Flows. AIAA*, AIAA 21st Aero. Sci Meeting, Reno, Nevada, USA.

Bushnell, D. M., (2003). *Aircraft drag reduction: a review. Jour. Aero. Eng.:25th Anneversary Collection, 217*, 1, 1 – 18.

Brasslow, A. L., (1999). *A History of Suction-Type Laminar-Flow Control with Emphasis on Flight Research. Monographs in Aerospace History, 13.*

Brunk, W. E., (1957). *Experimental Investigation of Transpiration Cooling for a Turbulent Boundary Layer in Subsonic Flow Using Air as a Coolant.*, TN 4091, Lewis Flight Propulsion Laboratory, Clevelend, Ohio. TN 4091: 36.

Benzi, R., (2010). *A short review on drag reduction by polymers in wall bounded turbulence. Physica D*, *239*, 1338 -– 1345.

Braslow, A., L., Collier, F., S., (1990). *Applied aspects of Laminar-Flow Technology, In viscous drag reduction in boundary layers*, Vol-123, *Prog. In Astronautics and Aeronautics.*

Brown, G., L. and Andrew, S., W., T., (1977). *Large structure in turbulent boundary layer, Phys. Fluids*, Vol-20, no-10, Part-2, pp. S243-S252.

Balakumar, B., J. And Adrian, A., J., (2007). *Large and very-large-scale motions in channel and boundary layer flows*, *Phil. Trans. R. Soc.*, Vol-365, pp. 665 – 681.

Bushnell, D., M. and Hefner, J., N. (eds), (1990a). *Viscous drag reduction in boundary layers*, Vol-123, *Prog. In Astronautics and Aeronautics, AIAA.*

Bushnell, D., M. and Hefner, J., N. (eds), (1990b). *Viscous drag reduction via surface mass injection*, Vol-123, *Prog. In Astronautics and Aeronautics, AIAA*, 1990.

Bushnell, D., M. and McGinley, C., B., (1989). *Turbulence control in wall flows, Ann. Rev. Fluid Mech.*, Vol-21, pp. 1 – 20.

Bradshaw, P., (1974). *Possible origin of Prandlts mixing length theory, Nature*, Vol.249, pp. 135 – 136.

Black, T., J. and Sarnecki, A., J., (1965). *The Turbulent Boundary Layer with Suction or Injection, A. R. C. Rep. No-3387.*

Corda, S., (2017) *Introduction to aerospace engineering with a flight test perspective. Wiley, ISBN:9781118953365* , 1–928.

Clauser, F., H., (1956). *The Turbulent boundary layer, Advances in App. Mech.*, Academic Press Inc., New York, Vol. 4, pp. 1 – 51.

Ching, C. Y., Djenidi, L. and R.A. Antonia,, (1994). *Low Reynolds Number Effects on the Inner Region of a Turbulent Boundary Layer, Developments in laser Techniques and Apllications to fluid mechanics, proccedings of 7th international symposium (Vol. 6)*, eds: Adrian, Durao, Durst, Madea, W.,Retrieved from `https://link.springer.com/content/pdf/10.1007%2F978-3-642-79965-5.pdf`, pp. 3 –15.

Corino, E., R. and Brodkey, R., S., (1969). *A visual study of turbulent shear flow. J. Fluid Mech., 37(1).*

Cai, Wei-Hua, Feng-Chen, Li, Zhang, Hong Na, Li, Xiao-Bin, Yu, B., Wei, Jinjia, Kawaguchi, Yasuo and Hishida, K. *Study on the characteristics of turbulent drag-reducing channel flow by particle image velocimetry combining with proper orthogonal decomposition analysis,Phys. Fluids, 11(21)*, 2009.

Crowe, T. C. Elger, D. F. Roberson, J. A. and Williams, B. C. *Engineering Fluid Mechanics, 8th ed, Wiley*, New York, 2005.

Choi, K.-S., (1961). *Theory of flow reattachment by tangential jet discharging against a strong adverse pressure gradient*, In: Boundary layer and flow control (G. V. Lachmann ed.) *London*, 209 – 231.

Choi, K.-S., (2001). *Turbulent Drag-Reduction Mechanisms: Strategies for Turbulence Management. Springer–Verlag, Vienna, 415* , 161–212.

Choi K.-S. Jukes T. and Whalley R., (2001). *Turbulent boundary-layer control with plasma actuators. Phil. Trans. R. Soc. A, 369*, 1443 — 1458.

Chin, C., Örlu, R., Monty, J., Hutchins, N., Ooi, A. and Schlatter, P., (2017). *Simulation of a Large-Eddy-Break-up Device (LEBU) in a Moderate Reynolds Number Turbulent Boundary Layer. Flow Turb. Combust., 98*, 445 – 460.

Corke, T. C., Guezennec, Y. and Nagib, H. M., (1981). *Modification in drag of turbulent boundary layers resulting from manipulation of large-scale structures. NASA CR 3444*, 1 – 25.

Carlier, J. and Stanislas, M., (2005) *Experimental study of eddy structures in a turbulent boundary layer using particle image velocimetry. J. Fluid Mech., 535*, 143 -– 188.

Coudert, S. and Schon, J., P., (2001). *Back-projection algorithm with misalignment corrections for 2D3C stereoscopic PIV. Meas. Sci. and Technol., 12*, 1371 – 1381.

Cuvier, C., (2012). *Active control of a separated turbulent boundary layer in adverse pressure gradient. PhD Thesis*, Ecole Centrale de Lille, 96.

Cuvier, C., Srinath, S., Stanislas, M., Foucaut, J., M., Laval, J., P., Kähler, C. J., Hain, R., Scharnowski, S., Schröder, A., Geisler, R., Agocs, J. , Röse, A., Willert, C., Klinner, J., Amili, O., Atkinson, C. and Soria, J., (2017). *Extensive characterisation of a high Reynolds number decelerating boundary layer using advanced optical metrology. Jour. Turb.*, 18:10, DOI: 10.1080/14685248.2017.1342827, 929 – 972.

Choi, H., Moin, P., Kim, J., (1994). *Active turbulence control for drag reduction in wall bounded flows. J. Fluid Mech.*, Vol-262, pp. 75 – 110.

Cantwell, J., B., (1981). *Organized motion in turbulent flow, Ann. Rev. Fluid Mech.*, Vol-13, pp. 457 – 515.

Craven, A., H., (1960). *Boundary layers with suction and injection-a review of published work on skin friction, College of Aeronautics-Cranfield*, rep-136, pp.1 – 43.

Catheral, D., Stewertson, K. and Williams, P., G., (1965). *Viscous Flow Past a Flat Plate With Uniform Injection, Proc. R. Soc. Lond. A*, vol-284, pp. 370 – 396.

Coles, D., (1956). *The Law of the wake in the turbulent boundary layer, J. Fluid Mech.*, Vol. 1, pp. 191 – 226.

Clauser, F., H., (1954). *Turbulent boundary layers at adverse pressure gradients, J. Aeronaut. Sci.*, Vol. 21, pp. 91 – 108.

Dantec Dynamics, (2014). *Measurement principle of Laser Doppler Anemometry*, `http://www.dantecdynamics.com/measurement-principles-of-lda`.

Dantec Dynamics (2006). *BSA Flow Software 4.0 Installation and User's Guide 0.1, Skovlunde, Denmark.* pp. 1 – 451.

Del Alamo, J. C., Jiménez, J., Zandonaze. P. and Moser. R. (2004). *Scaling of the energy spectra of turbulent channels. 30th Fluid Dynamics Conference*,Norfolk, Virginia, 39 – 56.

Daniello, R., J., Waterhouse, N., E., and Rothstein, J. P., (2009). *Drag reduction in turbulent flows over superhydrophobic surfaces, Phys. Fluids*, vol-21, 085103, pp. 1 – 9.

Dershin, H. and Leonard, C., (1967). *Direct measurement of skin friction on a porous flat plate with mass injection, AIAA*, Vol-11, pp. 1934 – 1967.

Durst, F., Kikura, H., Lekakis, I., Jovanović, J. and Ye, Q., (1996). *Wall shear stress determination from near-wall mean velocity data, Exp. Fluids*, Vol-20, pp. 417 – 428.

Dixon, W. J. and Massey, F. J. (1957). *Introduction to statistical analysis. MacGraw Hill.*

Eder, A., Durst, B. and Jordan, M. (2012). *Laser Doppler Velocimetry- Principle and application to turbulence measurements: In Optical measurements*, eds: Mayinger, F. and Feldmann, O., `https://doi.org/10.1017/CBO9781107415324.004` *Heat and mass transfer*, Vol-53.

Elbing, B., R., Winkel, E., S., Lay, K., A., Ceccio, S., L., Dowling, D., R. and Perlin, M., (2008). *Bubble induced skin friction drag reduction and the abrupt transitionto air layer drag reduction, J. Fluid Mech.*, Vol-612, pp. 201 – 206.

Eitel-Amor, G., (2014). *Simulation and validation of a spatially evolving turbulent boundary layer up to $Re_\theta$=8300. Int. Jour. of Heat and Fluid Flow, 47*, 57 – 69.

Fukagata, K. and Kasagi, N., P., K. (2006). *A theoretical prediction of friction drags reduction in turbulent flow by super hydrophobic surfaces, Phys. Fluids*, vol-18, 05703.1 – 4.

Falco, R., E., (1977). *Coherent motions in the outer region of turbulent boundary layers, Phys. Fluids*, Vol-20pp.S124 – S132.

Fukagata, K. and Kasagi, N., (2003). *Drag reduction in turbulent pipe flow with feedback control applied partially to wall, Int. J. Heat Fluid Flow*, Vol-24, pp. 480 – 490.

Fernholz, H., H. and Finley, P., J., (1996). *The incompressible zero-pressure-gradient turbulent boundary layer: an assessment of the data, Prog. Aerospace Sci.*, Vol-32, pp. 245 – 311.

Fukagata, K., Sugiyama, K. and Kasagi, N., (2009). *On the lower bound of net driving power in controlled duct flows, Physica D*, 1082 – 1086.

Fukagata, K., Iwamoto, K. and Kasagi, N., (2002). *Contribution of Reynolds stress distribution to the skin friction in wall-bounded flows. Phys. Fluids, 14*, L73 – L76.

Foucaut, J-M., Carlier, J. and Stanislas, M. (2004). *PIV optimization for the study of turbulent flow using spectral analysis. Meas. Sci. and Technol., 15*, 1046 – 1058.

Foucaut, J-M., Coudert, S., Stanislas, M. and Delville J. (2010). *Full 3D correlation tensor computed from double field stereoscopic PIV in a high Reynolds number turbulent boundary layer. Meas. Sci. and Technol., 50*, 839 – 846.

Foucaut, J-M., Coudert, S., Braud C. and Velte C. (2014). *Influence of light sheet separation on SPIV measurement in a large field spanwise plane. Exp. Fluids, 25*, 035304:1 – 035304:10.

Foucaut, J-M., Cuvier, C., Willert, C. and Soria, J. (2018). *Characterization of a high Reynolds number turbulent boundary layer by means of PIV. 5th International Conference on Experimental Fluid Mechanics (ICEFM)*, 372 – 377, 02.-04. Jul. 2018, Munich, Germany.

Gad-el-Haq, M., (2000). *Flow Control: Passive, Active, and Reactive Flow Management. Cambridge Uni. Press, New York, ISBN: 0-521-77006-8(hb).*

Gad-el-Haq, M., (1996) *Modern Developments in Flow Control (Review). American Soc. Mech. Eng., 49*, doi: 10.1115/1.3101931, 365–379.

Gad-El-Hak, M., (1990). *Control of low-speed airfoil aerodynamics. AIAA Jour.*, 28(9), 1537-1552.

Gad-el-Hak, M. and Bushnell, D., M., (1991). *Separation control: Review, ASME*, Vol-113/5, 1991.

Gad-el-Hak, (2012). *Control of low speed airfoil aerodynamics, AIAA*, Vol-28, no-9, pp.1537 – 1552.

Ganapathisubramani, B., Clemens, N. T. and Dolling, D. S. (2007). *Effects of upstream boundary layer on the unsteadiness of shock induced separation. J. Fluid Mech., 585*, 369 – 394.

García-Mayoral, R. and Jimenéz, J., (2011). *Drag reduction by riblets. Phil. Trans. R. Soc. A, 369*, 1412 -– 1427.

Grant, H., L., (1958). *The Large Eddies of Turbulent Motion. J. Fluid Mech., 4.*

Grin, V. T. (1967). *Experimental Study of Boundary-Layer Control by Blowing on a Flat Plate at M=2.5. Mekhanika Zhidkosti i Gaza*,2(6), 115 – 117.

Gregory, N., (1961). *Research on suction surfaces for laminar flow, Boundary Layer And Flow Control.* Eds: G. V. Lachmann, 924 – 960

Harun, Z., Monty, J. P. and Marusic, I., (2011). *The structure of zero, favorable and adverse pressure gradient turbulent boundary layers. 7th Int. Symp. Turb. Shear Flow Phenomena (TSFP), July*, 1 – 6.

Hough, G., R., (1980). *Viscous drag reduction, Prog. In Astronautics and Aeronautics: AIAA*, Vol-72.

Hoyt, J., W., (1990). *Drag reduction by polymers and surfactants, In viscous drag reduction in boundary layers, Prog. In Astronautics and Aeronautics*, Vol-123.

Hwang, D., P., (1997). *A proof of concept experiment for reducing skin friction by using a Micro-Blowing Technique, NASA Tech. Mem. 107315*, AIAA, vol-0546.

Hwang, D., P., (2002). *Experimental study of characteristics of micro-hole porous skins for turbulent skin friction reduction, ICAS Congress*, pp. 2101.1 – 7.

Hwang, D., P., (2003). *Review of research into concept of the microblowing technique for turbulent skin friction reduction, Prog. In Aerospace Science*, pp. 559 – 575.

Head, M., R. and Bandyopadhyay, P., R., (1981). *New aspects of turbulent boundary layer structure, J. Fluid Mech.*, Vol-107, no – 297.

Hutchins, N. and Marusic, I., (2007). *Large scale influences in near wall turbulence, Phil. Trans. R. Soc. A*, Vol-365, pp. 647 – 664.

Hamilton, J. M., Kim, J. and Waleffe, F., (1995). *Regeneration mechanisms of near-wall turbulence structures. J. Fluid Mech.*, Vol-287, pp. 317 – 348.

Hasanuzzaman, G. *Experimental investigation and CFD analysis of wind energy estimation considering building integrated ducts.* Unpublished master thesis, Brandenburg Technical University, Cottbus-Senftenberg, Germany, 2015.

Hassanuzzaman, G., Merbold, S., Motuz, V., Egbers, Ch., (2016). *Experimental Investigation of Turbulent Structures and their Control in Boundary Layer Flow, Fachtagung Experimentelle Strömungsmechanik*, 6. - 8. September 2016, pp. 64:1 – 6.

Hassanuzzaman, G., Merbold, S., Motuz, V., Egbers, Ch., Cuvier, C. and Foucaut, J-M. (2018). *Experimental investigation of active control inturbulent boundary layer using uniform blowing, 5th International Conference on Experimental Fluid Mechanics (ICEFM)*, 2. - 4. July 2018, pp. 1 – 6.

Hasanuzzaman, G., Merbold, S., Cuvier, C., Motuz, V., Foucaut, J.-M., and Egbers, Ch., (2020). *Experimental investigation of turbulent boundary layers at high Reynolds number with uniform blowing, part I: statistics, Jour. Turb.*, DOI:10.1080/14685248.2020.1740239, 21(3), pp. 129 – 165.

Hasanuzzaman, G., Merbold, S., Motuz, V. and Egbers, Ch., (2020b). *Enhanced outer peaks in turbulent boundary layer using uniform blowing at moderate Reynolds number, Jour. Turb., DOI:10.1080/14685248.2021.2014058..*

Hokenson, G., J., (1985). *Boundary Conditions for Flow Over permeable Surfaces, ASME*, vol.107, pp. 430 – 432.

Horn, M., Seitz, A. and Schneider, M., (2015). *Novel tailored skin single duct concept for HLFC fin application. 7TH EUROPEAN CONFERENCE FOR AERONAUTICS AND SPACE SCIENCES (EUCASS)*, 1 -– 11.

Hutchins, N. Nickels, T. B., Marusic, I. and Chong, M. S., (2009). *Hot-wire spatial resolution issues in wall-bounded turbulence. J. Fluid Mech., 635*, 103 -– 136.

Hoerner, S., F., (1965). *Fluid dynamic drag. New York.*

Hills, D., (2008). *The Airbus challenge. Flight Physics and Engineering div., EADS Engineering Europe*, Budapest.

Hussain, A., K., M., F. *Role of coherent structures in turbulent shear flows.*Proc.Indian Acad. Sci. *Engng. Sci.*, 4 1981, pp.129 – 175.

Hellsten, M., (2002). *Drag-reducing surfactants. Jour. Surfactants and Detergents, 5*, 65 – 70.

Hutchins, N. and Marusic, I., (2007). *Large-scale influences in near-wall turbulence. Phil. Trans. R. Soc., 365*, 647 -– 664.

Hutchins, N. and Marusic, I., (2007). *Evidence of very long meandering features in the logarithmic region of turbulent boundary layers. J. Fluid Mech., 579*, 1 -– 28.

Hwang, D., (2004). *Review of research into concept of the microblowing technique for turbulent skin friction reduction (Review). Prog. In Aerospace Sciences, 40*, 559 -– 575.

Herpin, S., Wong, C. Y., Stanislas, M. and Soria, J., (2008). *Stereoscopic PIV measurements of a turbulent boundary layer with a large spatial dynamic range. Exp. Fluids, 45*, 745 – 763.

Herpin, S., Stanislas, M., Foucaut, J-M. and Coudert, S., (2013). *Influence of the Reynolds number on the vortical structures in the logarithmic region of turbulent boundary layers. J. Fluid Mech., 716*, 5 -– 50.

International Air Transport Association, (2017). *Fact sheet fuel December 2017.* Retrieved from http://www.iata.org/pressroom/facts-figures/fact-sheets/Documents/factsheet-fuel.pdf,(accessed 8 January 2018).

International Maritime Organization, (2015). *Third International Maritime Organization Greenhouse Gas Study 2014: Executive summery and final report*, Micropress Printers, Suffolk, UK.

Jeong, J. and Hussain, F., (1995). *On the identification of a vortex, J. Fluid Mech.*, Vol-285, pp. 69 – 94.

Julien, H. I., Kays ,W. M., Moffat, R. J. (1969). *The Turbulent Boundary Layer on a Porous Plate: Experimental Study of the Effects of a Favorable Pressure Gradient. NASA*, R and M No-HMT4, 57.

Joslin, R. D., (1998). *Overview of laminar flow control (Review). NASA Tech. Rep.*, 1998-208707.

Jeong, J., and Hussain, F., (1995). *On the identification of a vortex, J. Fluid Mech.*, 285, 69 – 94.

Jiménez, J. and Pinelli, A., (1999). *The autonomous cycle of near wall turbulence, J. Fluid Mech.*, Vol-389, pp. 335 – 359.

Jiménez, J. and Moin, P., (1991). *The minimal flow unit in near-wall turbulence, J. Fluid Mech.*, vol-225, pp. 213 – 240.

Jeromin, L., O., F., (1970). *The status of Research in turbulent boundary layers with fluid injection, University of Cambridge Press.*

Kasagi, N., Hasegawa, N. and Fukagata, K., (2009). *Towards cost effective control of wall turbulence for skin friction drag reduction, European Turbulence Conference 12*, 2009.

Kito, M., Zanoun, E.-S., Jehring, L. and Egbers, C. (2006). *Laser-doppler-anemometry in turbulent boundary layers induced by different tripping devices compared with recent theories, Fachtagung "Lasermethoden in der Strömungsmesstechnik" GALA e.*, Vol-5-7, September 2006, Braunschweig, Germany, pp. 7.1 – 7.8.

Krishnan K. S. G. and Bertram O. *Assessment of a chamberless active HLFC system for the vertical tail plane of a mid-range transport aircraft, Deutscher Luft- und Raumfahrtkongress*, Doc ID 450080, 1-7, 2017.

Ko, Andy, Leifsson, L. T., Schetz, J. A., Mason, W. H., Grossman, B. and Haftka, R. T., (2003). *MDO (Multidisciplinary Design Optimization) of a Blended-Wing-Body Transport Aircraft with Distributed Propulsion. AIAA's 3rd Annual Aviation Technology, Integration, and Operations (ATIO) Technical Conference*, AIAA-2003-6732, 17 – 19 November, Denver, Colorado, 1 – 11.

Kim J., (2003). *Control of turbulent boundary layers. Phys. Fluids, 15*, 1093 — 1105.

Kim J. and Bewley T. R., (2007). *A Linear Systems Approach to Flow Control. Annu. Rev. Fluid Mech., 39*, 383 — 417.

Kasagi, N., Suzuki, Y. and Fukagata, K., (2009). *Microelectromechanical Systems–Based Feedback Control of Turbulence for Skin Friction Reduction. Annu. Rev. Fluid Mech., 41*, 231 – 251.

Kim, J. (2011). *Physics and control of wall turbulence for drag reduction. Phil. Trans. R. Soc. A., 369* , 1396 -– 1411.

Kornilov, V.I., (2005). *Reduction Of Turbulent Friction by Active and Passive Methods, Thermophys. Aeromech.*, Vol. 12, No. 2, P. 175 -– 196.

Kornilov, V. I. and Boiko, A. V. (2012). *Efficiency of Air Microblowing Through Microperforated Wall for Flat Plate Drag Reduction. American Institute of Aeronautics and Astronautics, 50*(3), 724 -– 732.

Kornilov, V. I., (2015). *Current state and prospects of researches on the control of turbulent boundary layer by air blowing (Review). Prog. in Aerospace Sciences, 76*(2), 1 -– 23.

Kostas, J., Foucaut, J-M. and Stanislas, M. (2005). *Application of Double SPIV on the Near Wall Turbulence Structure of an Adverse Pressure Gradient Turbulent Boundary Layer. Proceedings of the 6th International Symposium on PIV, Pasadana*, CA.

Kähler, C. J., Scharnowski, S., and Cierpka, C., (2016). *Highly resolved experimental results of the separated flow in a channel with streamwise periodic constrictions. J. Fluid Mech., 796*, 257 – 284.doi:10.1017/jfm.2016.250

Kametani, J. and Fukagata, K., (2011). *Direct numerical simulation of spatially developing turbulent boundary layers with uniform blowing or suction. J. Fluid Mech., 681*, 154 – 172.

Kametani, J., Fukagata, K., Örlü, R. and Schlatter, P. (2015). *Effect of uniform blowing/suction in a turbulent boundary layer at moderate Reynolds number. Int. Jour. Heat and Fluid Flow, 55*, 132 – 142.

Kametani, J., Fukagata, K., Örlü, R. and Schlatter, P. (2016). *Drag reduction in spatially developing turbulent boundary layers by spatially intermittent blowing at constant mass-flux. Jour. Turb., 17(10)*, 913 – 927.

Kim, J., Moin, P. And Moser, R., (1987). *Turbulence statistics in fully developed channel flow at low Reynolds number, J. Fluid Mech.*, Vol-177, pp. 133 – 166.

Kasagi, N., Suzuki, Y. and Fukagata, K., (2009). *Microelectromechanical systems based feedback control of turbulence for skin friction reduction, Ann. Rev. Fluid. Mech.*, 41-231-51.

Kline, S., J., Reynolds, W., C., Schraub, F., A., and Rundstadtler, P., W., (1967). *The structure of turbulent boundary layers, J. Fluid Mech.*, Vol-30, pp. 741 – 773.

Kovasznay, L., S., G., Kibens, V. and Blackwelder, R., F., (1970). *Large scale motion in the intermittent region of a turbulent boundary layer, J. Fluid Mech.*, Vol-41, part-2, pp. 283 – 325.

Klebanoff, P., S., (1954). *Characteristics of turbulence in a boundary layer with zero pressure gradient, NACA Tech. Rep. 1247.*

Klebanoff, P., S., (1950). *Some features of artificially thickened fully developed turbulent boundary layers with zero pressure gradient, NACA Tech. Rep. 2475.*

Kim, J., (1992). *Study of turbulence structure through numerical simulations: the perspective of drag and reduction, AGARD Rep. 786.*

Kasagi, N. and Shikazono, N., (1995). *Contribution of direct numerical simulation to understanding and modelling turbulent transport, Proc. R. Soc. London*, Vol-451, pp. 257 – 292.
Lumley JL (1970) Stochastic Tools in Turbulence. Academic Press, New York

Lumley J., L., (1970). *Stochastic Tools in Turbulence, Academic press, New York.*

Lee. M., K., Eckelman, L., D. and Hlanratty, T., J., (1974). *Identification of turbulent wall eddies through the phase relation of the components of the fluctuating velocity gradient, J. Fluid Mech.*, Vol-66, pp. 17.

Leonardi,S. and Orlandi,P. and Antonia,R. A. , (2007). *Properties of d- and k-type roughness in a turbulent channel flow,* `https://doi.org/10.1063/1.2821908`,*Phys. Fluids*, Vol-19 (12), pp. 125101.1 – 6.

Lissaman, P. B. S., (1983). *Low Reynolds number aerofoils. Ann. Rev. Fluid Mech.*, doi: 10.1146/annurev.fl.15.010183.001255, **15**: 223 – 239.

Liu, C. K., Kline, S. J. and Johnston, J. P., (1966). *An experimental study of turbulent boundary layers on rough walls. Thermoscience Div., Mech. Eng. Dept., Stanford Univ.,, Rep. MD-15.*

Legner, H., H., (1984). *A simple model for gas bubble drag reduction. Phys. Fluids, 27,* 2788 – 2790.

Lynch, F., T., Klinge, M., D., (1991). *Some practical aspects of viscous drag reduction concepts, SAE Conf. Paper*, vol-912129.

Lumley, J. and Blossey, P., (1998). *Control of Turbulence, Ann. Rev. Fluid Mech.*, Vol-30, pp. 311 – 327.

Leadon, B., M., (1961). *Comments on A sublayer theory for fluid injection, Jour. Aero. Sci.*, Vol-28, no-9, pp.826 – 827.

Monty JP, Stewert JA, Williams RC and Chong MS (2007), Large-scale features in turbulent pipe and channel flows, *J. Fluid Mech.*, 589: 147 – 156

McLaughlin, D. K. and Tiederman, W. G., (1973). *Biasing correction for individual realisation of laser anemometer measurements in turbulent flow. Phys. Fluids, 16,* 2082 – 2088.

Monkewitz, P. A., Chauhan, K. A. and Nagib, H. M., (2007). *Self-consistent high-Reynolds-number asymptotics for zero-pressure-gradient turbulent boundary layers.. Phys. Fluids, 19(11)*, 115101.

Millikan C., M., (1938). *A critical discussion of turbulent flows in channels and circular tubes. Proceedings fifth int. congress appl. mech.*, Cambridge University Press, 386 – 392.

Marusic, I., Talluru, K., M. and Hutchins, N. (2014). *Controlling the Large-Scale Motions in a Turbulent Boundary Layer. Y. Zhou et al. (eds.), Fluid-Structure-Sound Interactions and Control*, Springer-Verlag Berlin Heidelberg 2014, *27*, 17 – 26.

Mathis, R., Hutchins, N. and Marusic, I., (2009). *Large-scale amplitude modulation of the small-scale structures in turbulent boundary layers. Jour. Fluid Mech.*, *628*, 311 – 337.

Mathis, R., Marusic I. and Hutchins N. (2010). *Predictive model for wall-bounded turbulent flow.. Science*, *329*, 193 -– 196.

Merkle, C., L. and Deutsch, S., (2010). *Microbubble drag reduction. Frontiers in Experimental Fluid Mechanics*, Editor: M. Gad-el-Hak, *46*, 291 – 335.

Mahfoze, O. and Leizet S., (2017). *Skin-friction drag reduction in a channel flow with streamwise-aligned plasma actuators. Int. Jour. Heat and Fluid Flow*, *66*, 83 — 94.

Marusic, I., Mathis, R. and Hutchins, N., (2010a) *High Reynolds number effect in wall turbulence. Int. J. Heat Fluid Flow*, *31(3)*, 418 -– 428.

Marusic, I., McKeon, B. J., Monkewitz, P. A., Nagib, H. M., Smits, A. J., (2010b). *Wall-bounded turbulent flows at high Reynolds numbers: Recent advances and key issues. Phys. Fluids*, *22*, 065103.1 -– 24.

Mehregany, M., DeAnna, R., G. and Reshotko, E., (1996). *Microelectromechanical systems for aerodynamics applications*, *ARL Tech. Rep.1113-NASA Tech. Mem.*, 107320.

Min, T. and Kim, J., (2004). *Effects of Hydrophobic surface on skin friction drag*, *Phys. Fluids*, vol-16, no-7, pp. 55 – 58.

Martell, M., B., Perot J., B., and Rothstein, J., P., (2009). *Direct numerical simulations of turbulent flows over superhydrophobic surfaces*, *J. Fluid Mech.*, vol. 620, pp. 31 – 41.

Moin, P., (1993). *Active turbulence control in wall bounded flows using Direct Numerical Simulation*, *AFOSR/NA-Stamford University Tech. Rep.*.

Marusic, I. and Heuer, W., (2007). *Reynolds number invariance of the structure inclination angle in wall turbulence*, *Phys. Rev. Letter*, Vol- 99, 114504.

Motuz, V. (2014). *Gleichmäßiges Mikro-Ausblasen zur Beeinflussung einer turbulenten Grenzschicht*, *Ph.D. Thesis*, BTU, http://nbn-resolving.de/urn:nbn:de:kobv:co1-opus4-31242 (Advisors: Ch. Egbers, U. Rist, V.I. Kornilov).

Mickley, H., S., Ross, R., C., Squyers, A., L. and Stewert, W., E., (1954). *Heat, Mass and Momentum transfer for Flow over a Flat Plate Turbulent Boundary layer*, *NACA Tech. Rep.3208, MIT.*

Mickley, H., S. and Davis, R., S., (1957). *Momentum transfer for flow over a flat plate with blowing*, *NACA Tech. Note. 4017.*

McQuaid, J., (1968). *Experiments in Incompressible Turbulent Boundary Layers with Distributed Injection*, *A. R. C. Rep. No-3549.*

Menon, S. (2003). *Large-eddy/lattice Boltzmann simulations of micro-blowing strategies for subsonic and supersonic drag control. NASA/CR—2003-212196.*

Menon, S. (2003). *Micro Blowing Simulations Using a Coupled Finite-Volume Lattice-Boltzmann LES Approach. CCL Report 2005-009: 62.*

Nagib, H., M., Chauhan, K., A., and Monkewitz, P., A., (2007). *Approach to an asymptotic state for zero pressure gradient turbulent boundary layers. Phil. Trans. Ro. Soc.*, 365(1852), pp. 755 – 770.

NATO, (1985). *Aircraft drag prediction and reduction, AGARD-NASA-Von Karman Institute Report no-723.*

Örlü, R. and Schlatter, P. (2013).*Experiments and simulations for zero pressure gradient turbulent boundary layers at moderate Reynolds numbers.Exp. Fluids, 43* , 665 -– 681.

Örlü, R. and Schlatter, P., (2015). *Comparison of experiments and simulations for zero pressure gradient turbulent boundary layers at moderate Reynolds numbers. Exp. Fluids, 43*:1547, 1 – 21.

Österlund, J. M., Johansson. A., V., Nagib. H. M. and Hites. M. H. (1999). *Experimental studies of zero pressure-gradient turbulent boundary layer flow. Ph. D. Thesis*,Royal Institute of Technolohy (KTH), Sweden.

Österlund, J. M., Johansson. A., V., Nagib. H. M. and Hites. M. H. (1999). *Mean-flow characteristics of High Reynolds Number Turbulent Boundary Layers from Two Facilities. 30th Fluid Dynamics Conference*,Norfolk, Virginia, 39 – 56.

Pope, S., B., (2000). *Turbulent Flows, Cornell University Press.*

Piomelli, U., Moin, P. and Ferziger, J., (1989). *Large Eddy Simulation of the flow in a transpired channel, J. Thermophysics, AIAA*, vol-1, 124 – 128.

Panton, R., L., (2001). *Overview of the self-sustaining mechanisms of wall turbulence. Prog. Aero. Sci., 37*, 341 – 383.

Prasad, A., K., (2000). *Stereoscopic particle image velocimetry. Exp. Fluids, 29*, 103 – 116.

Prasad, A., K. and Jensen, K., (1995). *Scheimpflug stereocamera for particle image velocimetry in liquid flows. Appl. Optics, 34*, No-30, 7092 – 7099.

Plas, A. P., Sargeant, M. A., Madani, V., Crichton, D., Greitzer, E. M., Hynes, T. P. and Hall, C. A., (2007). *Performance of a Boundary Layer Ingesting (BLI)Propulsion System. 45th AIAA Aerospace Sciences Meeting and Exhibit*, AIAA 2007-450, 8 – 11 January, Reno, Nevada, 1 – 21.

Pohlhausen, K., (1921). *Zur Näherungsweisen Integration der DifferenzialgJeichung der Lainaren Grenzschicht, Z. Angew. Math. M. Echo.*, Vol-1, 252 – 68.

Prandlt, L., (1927). *Proceedings of the second Congress of applied Mechanics*, Zurich, 1926, *translated from Aeronautical Research Council paper, Ae. Tech.*

Prandlt, L., (1935). *The mechanics of viscous fluids, in: Durand, W., F., Aerodynamic theory, (iii)*, pp. 34 – 208.

Tiederman, W. G., and Reischman, M. M. (1975). *Laser-Doppler anemometer measurements in drag-reducing channel flows. jour. fluid mech.*, *70*, 369 – 392.

Rheinboldt, W., (1956). *Zur Berechnung stationärer Grenzschichten bei kontinuierlicher Absaugung mit unstetig veränderlicher Absaugegeschwindigkeit. J. Rat. Mech. Analysis*, *5*, 539 -- 596.

Rothstein, J. P., (2010). *Slip on Super hydrophobic surfaces (Review). Annu. Rev. Fluid Mech.*, *42*, 89 -- 109.

Reneaux, J., (2004). *Overview on drag reduction technologies for civil transport aircraft. Meth. Appl. Sci. Eng.*, ECCOMAS 2004, Jyväskylä, 24—28 July 2004, 1 – 18.

Robinson, S., K., (1991). *Coherent motions in the turbulent boundary layers, Ann. Rev. Fluid Mech.*, Vol-23, pp. 601 – 639.

Rubesin, W., M., (1954). *An analytical estimation of the effect of transpiration cooling on the heat transfer and skin friction characteristics of a compressible turbulent boundary layer, NACA Tech. Note. 3341.*

Richardson, S., A., (1971). *A Model for the Boundary Condition of a Porous Material*, Part-2, *Jour. Fluid Mech.*, vol.49, pp.327 – 336.

Rotta, R., J., (1953). *On the theory of the turbulent boundary layer, NACA Tech. Mem. 1344.*

Rotta, J. C., (1950). *Über die Theorie der Turbulenten Grenzschichten. Mitt. M.P.I.. Ström. Forschung Nr 1 (also available as NACA TM 1344).*

Rotta, R., J., (1970). *Control of turbulent boundary layers by uniform injection and suction of fluid, ICAS Paper no-70-10.*

Sumitani, Y. and Kasagi, N., (1995). *Direct Numerical Simulation of turbulent transport with uniform wall injection and suction, AIAA*, vol-33, 1220 – 1228.

Sutherland, W., (1893). *The viscosity of gases and molecular force. Philosophical Magazine*, 5(36), pp.507 – 531.

Schlichting, H., (1960). *Boundary layer theory. McGraw-Hill*, New York.

Smits, A. J., McKeon B. J., Marusic, I., (2011). *High Reynolds number wall turbulence. Annu. Rev. Fluid Mech.*, *43*, 353 – 375.

Spallart P. R. and McLean J. D., (2011). *Drag reduction: enticing turbulence, and then an industry (Review). Phil. Trans. R. Soc. A, 369*, 1556 -- 1569.

Stroh A., Frohnapfel B., Schlatter P. and Hasegawa Y., (2015). *A comparison of opposition control in turbulent boundary layer and turbulent channel flow. Phys. Fluids, 27*, 075101-1 — 075101-14.

Stroh A., Hasegawa Y., Schlatter P. and Frohnapfel B. Stroh A., Frohnapfel B., Schlatter P. and Hasegawa Y., (2016). *Global effect of local skin friction drag reduction in spatially developing turbulent boundary layer. Jour. Fluid Mech., 805*, 303 – 3021.

Scarano, F., (2002). *Iterative image deformation methods in PIV. Meas. Sci. and Technol., 13*, R1 – R19.

Soria, J., Cater, J. and Kostas, J., (2000). *High resolution multigrid cross-correlation digital PIV measurements of a turbulent starting jet using half frame image shift film recording. Laser and Optics Tech., 31*, 1, 3 – 12.

Soria, J., (1996). *An investigation of the near wake of a circular cylinder using a video-based digital cross-correlation particle image velocimetry technique. Laser and Optics Tech., 31*, 12, 221 – 233.

Soloff, S. M., Adrian, R. J. and Liu, Z. C. (1997). *Distortion compensation for generalized stereoscopic particle image velocimetry. Meas. Sci. and Technol., 8*, 1441 – 1454.

Spalart, P.R., Strelets, M. and Travin, A., (2001). *Direct numerical simulation of large-eddy-break-up devices in a boundary layer. Phil. Trans. R. Soc. A, 27, 5*, 902 — 910.

Smith, L. H., (1993). *Wake Ingestion Propulsion Benefit. Jour. Propulsion Power, 9*, No.1, 74 – 82.

Sciacchitano, A. and Wieneke, B. (2016). *PIV uncertainty propagation. Meas. Sci. Technol., 27*, 1 – 16,(doi:10.1088/0957-0233/27/8/084006).

Srinath, S., Vassilicos, J., C., Cuvier, C., Laval, J.-P., Stanislas, M. and Foucaut, J.-M. (2018). *Attached flow structure and streamwise energy spectra in a turbulent boundary layer. Phys. Rev. E, 97(5)*, 1 – 14.

Schlichtling, H., (1942a). *The boundary layer on a flat plate under conditions of suction and air injection (Die Grenzschicht an der ebenen Platte mit Absaugung und Ausblassen), Luftfahrtforschung*, vol.19, pp.293 – 301.

Schlichtling, H., (1942b). *Die Grenzschicht mit Absaugung und Ausblassen, Luftfahrtforschung*, vol.19, pp.179 – 181.

Spalart, P., (1988). *Direct numerical simulation of a turbulent boundary layer up to* $Re_\theta$ *=1410, J. Fluid Mech.*, Vol-187, pp. 61 – 98.

Schlatter, P., Örlu, R., Li, Q., Brethouwer, G., Fransson, J., H., M., Johansson, A. ,V., Alfredsson, P. H. and Henningson, D.S., (2009b). *Turbulent boundary layers up to* $Re_\theta = 2500$ *studied through simulation and experiment. Phys. Fluids*, Vol-21, 51702.

Schlatter, P., Örlu, R., Li, Q., Brethouwer, G., Johansson, A. ,V., and Henningson, D.S., (2009). *High Reynolds number turbulent boundary layers studied by numerical simulation, Bulletin American Phys. Soc., APS Phys.*, Vol-54, pp.59, .

Schlatter, P., Li, Q., Brethouwer, G., Johansson, A., V., and Henningson, D., S., (2010a). *Simulations of spatially evolving turbulent boundary layers up to* $Re_\theta = 4300$*,Int. J. Heat Fluid Flow*, Vol-31, pp. 251 – 261.

Schlatter, P., and Örlu, R., (2010b). *Assessment of direct numerical simulation data of turbulent boundary layers, J. Fluid Mech.*, Vol-659, pp. 116 – 126.

Schlatter, P. and Örlu, R., (2012). *Turbulent boundary layers at moderate Reynolds numbers: inflow length and tripping effects, J. Fluid Mech.*, Vol-710, pp. 5 – 34.

Schlatter, P., Li, Q., Örlü, R., Hussain, F., Henningson, D.S., (2014). *On the near-wall vortical structures at moderate Reynolds numbers, Eur. Jour. Mech. B/Fluids*, vol. 48 pp. 75 – 93.

Smits, A., J., Matheson, N. and Joubert, P., N., (1983). *Low Reynolds number TBL in zero and favorable pressure gradients, Jour. of Ship Research*, Vol-27, no-3, pp. 147 – 157.

Stevenson, T., N., (1963). *A law of the wall for turbulent boundary layers with suction or injection, College or Aero. Cranfield Rep. No. 166.*

Simpson, R., L., Moffat, R., J. and Kays, W., M., (1968). *The turbulent boundary layer on a porous plate: experimental skin friction with variable injection and suction, Int. J. Heat Mass Transfer*, Vol. 12, pp. 771 – 789.

Springer (2007). *Handbook of Experimental Fluid Mechanics. Eds. Tropea, C., Yarin, A. L. and Foss, J. F..*

Tamano, S., Itoh, M., Kato, K. and Yokota, K., (2010). *Turbulent drag reduction in nonionic surfactant solutions. Phys. Fluids, 22*, 055102.1 – 12.

Tang,, Z., Jiang, N., Zheng, X. and Wu, Y., (2019). *Local dynamic perturbation effects on amplitude modulation in turbulent boundary layer flow based on triple decomposition. Phys. Fluids, 31(2)*, 025120.1 – 14.

Townes H., W. and Sabersky, R., H., (1966). *Experiments on the flow over rough surface, Int. J. Heat Mass Transfer*, vol-9, pp. 729 – 738.

Townsend, A., A., (1976). *The structure of Turbulent Shear Flow, 2nd ed, Cambridge University press*, Cambridge.

Theodorsen, T., (1952). *Mechanism of Turbulence, In Proc. 2nd Midwest. Conf. Fluid Mech.*, Ohio State University, pp. 1 – 19.

Turcotte, D., L., (1960). *A sublayer theory for fluid injection into the incompressible turbulent boundary layer, Jour. Aero. Sci.*, Vol-27, no-9, pp.675 – 678.

Taylor, G., I., (1971). *A Model for the Boundary Condition of a Porous Material*, Part-1, *Jour. Fluid Mech.*, vol.49, pp.319 – 336.

Tennekes, H., (1965). *Similarity laws for turbulent boundary layers with suction or injection, J. Fluid Mech.*, Vol. 21, pp. 689 – 703.

Tennekes, H. and Lumley, J., L. (1972). *A first course in turbulence, MIT Press*, Cambridge, Massachusetts.

Thomas, F. (1965). *On boundary-layer control for increasing lift by blowing. AIAA*, 3(5) , 967 – 968.

Van Driest, E. R., (1956). *On turbulent flow near a wall. AIAA, 23*:11, 1007 – 1011, 1036.

Vallikivi, M., Hultmark, M. and Smits A. J., (2015a). *Turbulent boundary layer statistics at very high Reynolds number. J. Fluid Mech., 779*, 371 – 389.

Vallikivi, M., Ganapathisubramani, B. and Smits A. J., (2015b). *Spectral scaling in boundary layers and pipes at very high Reynolds numbers. J. Fluid Mech., 771*, 303 – 326.

Vukoslavčević, P. Wallace, J., M. and Bailnt, J., L., (1991). *Viscous drag reductions using streamwise aligned riblets. AIAA, 30*, 1119 – 1125.

Wallace, J., M., (2016). *Quadrant analysis in turbulence research: History and evolution. Ann. Rev. Fluid Mech., 48*, 131 – 158.

Warsop C. Current status and prospects for flow control. *Aerodynamic drag reduction technologies*, Springer-Verlag Berlin Heidelberg GmbH, **999**: 269-277, 2000.

Wei, T., Fife, P., Klewicki, J., and McMurtry, P., (2005). *Properties of the mean momentum balance in turbulent boundary layer, pipe and channel flows. Jour. Fluid Mech., 522*, 303 – 327.

Wood, R. M., (2004). *Impact of Advanced Aerodynamic Technology on Transportation Energy Consumption. SAE International 2004*, Technical Paper, *2004-01-1306.*.

Willert, C. J., (2015). *Experimental Evidence of Near-Wall Reverse Flow Events in a Zero Pressure Gradient Turbulent Boundary Layer. Phys. Fluid Dyn.*, 01 – 08 .

Wieneke, B. (2005). *Stereo-PIV using self-calibration on particle images. Exp. Fluids, 39*, 267 – 280.

Westerweel, J., Dabiri, D. and Gharib, M., (2000). *The effect of a discrete window offset on the accuracy of cross-correlation analysis of digital PIV recordings. Exp. Fluids, 23*, 1, 20 – 28.

Willert, C. A. and Gharib, M. (1991). *Digital particle image velocimetry. Exp. Fluids, 10*, 181 – 193.

White, C. M. and Mungal, M. G., (2008). *Mechanics and Prediction of Turbulent Drag Reduction with Polymer Additives (Review). Annu. Rev. Fluid Mech., 40*, 235 – 256.

Walsh M.J. and Lindemann A.M., (1984). *Optimization and application of riblets for Turbulent Drag Reduction. AI AA-84-0347, 22nd Aerospace Sciences Meeting,*Reno, Nevada, USA, *9-12 January*, 1 — 10.

Wilkinson, S. P., Anders, J. B., Lazos, B. S. and Bushnell, D. M., (1988). *Turbulent drag reduction research at NASA Langley: Progress and Plans. Int. J. Heat and Fluid Flow, 9, 3*, 266 — 277.

Watanabe, S., Mamori, H. and Fukagata, K., (2017). *Drag-reducing performance of obliquely aligned super hydrophobic surfaces in turbulent channel flow, Fluid Dyn. Research* vol-49, pp. 1 – 20.

Wu, X., and Moin, P., (2009). *Direct numerical simulation of turbulence in a nominally zero-pressure- gradient flat plate boundary layer, J. Fluid Mech.*, Vol-630, pp. 5 – 41.

Waleffe, F., (1997). *On a self-sustaining process in shear flows, Phys. Fluids*, vol- 9, pp. 883 – 900.

White, M., F., (1974). *Viscous Fluid Flow, 2nd ed., McGraw Hill*, ISBN 0-07-100995-7.

Winter, K., G., (1977). *An outline of the techniques available for the measurement of skin friction in turbulent boundary layers, Prog. Aerospace Sci.*, Vol. 4, pp. 1 – 57.

Xu, cx., Bing-Qing, D., and Wei-Xi, H. and Guixiang, C., (2013). *Coherent structures in wall turbulence and mechanism for drag reduction control. Sci. China Phys., Mech. Astr., 56(6)*, 1053 – 1061.

Yoshioka, S. and Alfredsson, H., (2006). *Control of turbulent boundary layers by uniform wall suction and blowing, Govindarajan, R. (eds), 6th IUTAM Symposium on Laminar turbulent transition*, pp. 437 – 442.

Zhu, W., van Hout, R. and Katz, J. (2007). *On the flow structure and turbulence during sweep and ejection events in a wind-tunnel model canopy. Boundary-Layer Meteorol., 124*, 205 – 233, `https://doi.org/10.1007/s10546-007-9174-9`

Zagarola, M. and Smits, A., J., (1998). *Mean-flow scaling of turbulent pipe flow. Jour. Fluid Mech., 373*, 33 — 79.

Zhou, J., Adrian, R., J., Balachandar, S. And Kendall, T., M., (1999). *Mechanisms for generating coherent packets of hairpin vortices in channel flow*, *J. Fluid Mech.*, Vol-387, pp. 353 – 396.

Zanoun, E.-S., Kito, M., and Egbers, Ch. (2009). *A study on flow transition and development in Circular and Rectangular Ducts*, *J. Fluids Eng.*, Vol-131, 061204.

Zanoun, E. -S. (2010). *Mean flow scaling along smooth and rough wall boundary layers*, *Thermophys. Aeromech.*, Vol-17(1), pp. 21–38.

Zanoun, E.-S., Jehring, L. and Egbers, Ch., (2014). *Three measuring techniques for assessing the mean wall skin friction in wall-bounded flows*, *Thermophys. Aeromech.*, Vol-21(2), pp. 179 – 190.

www.ingramcontent.com/pod-product-compliance
Ingram Content Group UK Ltd.
Pitfield, Milton Keynes, MK11 3LW, UK
UKHW022001190726
13853UKWH00004B/1658

9 783736 975583